Fodil Hamzaoui
Abdelakader Chouaih

Electron Charge Density in Organic Coumpounds

Fodil Hamzaoui
Abdelakader Chouaih

Electron Charge Density in Organic Coumpounds

Non Linear Opttical Properties

LAP LAMBERT Academic Publishing

Impressum / Imprint
Bibliografische Information der Deutschen Nationalbibliothek: Die Deutsche Nationalbibliothek verzeichnet diese Publikation in der Deutschen Nationalbibliografie; detaillierte bibliografische Daten sind im Internet über http://dnb.d-nb.de abrufbar.
Alle in diesem Buch genannten Marken und Produktnamen unterliegen warenzeichen-, marken- oder patentrechtlichem Schutz bzw. sind Warenzeichen oder eingetragene Warenzeichen der jeweiligen Inhaber. Die Wiedergabe von Marken, Produktnamen, Gebrauchsnamen, Handelsnamen, Warenbezeichnungen u.s.w. in diesem Werk berechtigt auch ohne besondere Kennzeichnung nicht zu der Annahme, dass solche Namen im Sinne der Warenzeichen- und Markenschutzgesetzgebung als frei zu betrachten wären und daher von jedermann benutzt werden dürften.

Bibliographic information published by the Deutsche Nationalbibliothek: The Deutsche Nationalbibliothek lists this publication in the Deutsche Nationalbibliografie; detailed bibliographic data are available in the Internet at http://dnb.d-nb.de.
Any brand names and product names mentioned in this book are subject to trademark, brand or patent protection and are trademarks or registered trademarks of their respective holders. The use of brand names, product names, common names, trade names, product descriptions etc. even without a particular marking in this work is in no way to be construed to mean that such names may be regarded as unrestricted in respect of trademark and brand protection legislation and could thus be used by anyone.

Coverbild / Cover image: www.ingimage.com

Verlag / Publisher:
LAP LAMBERT Academic Publishing
ist ein Imprint der / is a trademark of
OmniScriptum GmbH & Co. KG
Heinrich-Böcking-Str. 6-8, 66121 Saarbrücken, Deutschland / Germany
Email: info@lap-publishing.com

Herstellung: siehe letzte Seite /
Printed at: see last page
ISBN: 978-3-659-71033-9

ELECTRON CHARGE DENSITY ANALYSIS

IN ORGANIC COMPOUNDS

Fodil HAMZAOUI and Abdelkader CHOUAIH
University of Mostaganem – Algeria

Symmary

Introduction

In this last decade our work was focused on studying the charge density analysis of a variety of organic compounds used in different areas such us telecommunication, solar energy or biology. It is known that some physical properties such us the non-linear optical response and the biological activity of the studied compounds are directly linked to their electronic molecular structure.

The knowledge of the electron density function derived from x-ray diffraction studies combined to theoretical ab-initio calculation can lead to the rightful description of the molecular electrostatic properties such the dipole moment and the electrostatic potential around the molecules.

In all the presented articles, the crystallographic investigation is based on the exploitation of high resolution x-ray experiment data. The description of the electron density of the studied molecules is based on the Coppens-Hansen multipolar model which describes the crystal electron density as a superposition of aspherical pseudo-atoms each modelled on a multipole expansion:

$$\rho_k(\vec{r}) = P_{k,c}\rho_{k,c}(\vec{r}) + P_{k,v}\kappa^3\rho_{k,v}(\kappa,\vec{r}) + \sum_{l=0}^{4}\kappa'^3 R_{k,l}(\kappa',\vec{r}) \times \sum_{m=-\ell}^{+\ell} P_{k,lm\pm}Y_{lm\pm}\left(\frac{\vec{r}}{|\vec{r}|}\right)$$

Here ρ_c and ρ_v are spherically averaged Hartree-Fock core and valence densities, with ρ_v normalized to one electron. The Slater type radial functions $R_l(r) = N_\ell r^{n_\ell}\exp(-\kappa'\xi_\ell r)$ modulate the spherical harmonic angular functions $Y_{lm\pm}$. The sum over m in Equation (1) includes $\pm l$, so that for each one, $2l +1$ functions are included. The adjustable variables are the valence shell contraction-expansion parameters (κ, κ') and the population parameters (P_v, $P_{lm\pm}$). The aspheric charge density was described at the octupole ($l = 3$) level for all non-hydrogen atoms and at the dipole levels (l = 1 or 2) for hydrogen atoms not involved and involved in strong H-bonds, respectively. The core and valence scattering factors came from International Tables for X-ray Crystallography (1999). Charge densities of all hydrogen were considered to have cylindrical symmetry along the corresponding hydrogen-heavy atom bond. A molecular electro-neutrality constraint was applied in all refinements.

The theoretical calculations are performed using the Density Functional Theory at B3LYP (Becke's three parameter hybrid functional using the correlation functional of Lee, Yang, and Parr, which includes both local and non-local terms correlation functional) methods at 6-31G* level. In order to perform this computational work, we used the Gaussian 09 program package.The Highest Occupied Molecular Orbital(HOMO–Lowest Unoccupied Molecular Orbital (LUMO) analysis has been carried out to explain the charge transfer place within the molecule. The chemical hardness and chemical potential are also calculated using the HOMO and LUMO.

The visualization of the electron charge and the electrostatic potential of the molecule were obtained using the Molden program where the Molden is a package for displaying molecular density from the *ab initio* package Gaussian.

All the presented articles provide a comparison of the experimental and theoretical results. The molecular dipole moment and the electrostatic potential around the studied molecules are systematically given.

We hope that this work can help young researchers beginning in electron charge density investigations by discovering some physical properties related to the used method when investigating organic compounds.

THE MOLECULAR LINEAR POLARIZABILITY FROM X-RAY DIFFRACTION STUDY THE CASE OF 3-METHYL-4-NITROPYRIDINE-1-OXYDE (POM).

Fodil Hamzaoui, Abdelwahab Zanoun and Gérard Vergoten

Abstract

The main purpose of this work, besides x-ray diffraction electronic density determination of a nonlinear molecular crystal, is to revisit the earlier Robinson model (1967) which proposed to derive the nonlinear optical properties of molecule from its ground state charge distribution. The present study is mainly based on the ground state charge distribution inferred from X-ray diffraction data and the application of the Unsöld approximation through the Robinson model. This model has been applied to the POM a prototypical nonlinear organic crystal, whereby relations between polarisabilities and multipolar moments of the electronic charge distribution have been evidenced.

The experimental electronic density analysis has been carried out by use aspherical model of Hansen and Coppens (1978). The electrostatic moments were then estimated by direct integration methods using both discreet and fuzzy boundary space partitioning. The electrostatic moments were also inferred from the charges derived from a semi-empirical calculation implemented in the electronic part of MOPAC (MNDO, PM3 Hamiltoniens). The results of this so called hereafter 'Point Charge Model' show relatively good agreements with those derived from X-ray electron density distribution except however for the component along X. The validity of the Unsöld approximation has been checked by comparing the values of the linear α property estimated from the Finite Field method with those deduced from Robinson model using the ground state moments of the point charge model. Comparison of the results obtained on the free molecule using the Point Charge Model with those derived from the experimental charge distribution seems to reveal interesting information as to the influence of the crystal field effects on the molecular properties.

Introduction.

In the optical domain, a molecular engineering approach based on such principles has led to development of organic crystals displaying comparable or even better non linear optical properties than known inorganic compounds (Chemla and Zyss, 1987). It has been evidenced that molecules of those materials with donor and acceptor substituted at parapositions (such as para-nitroaniline or nitropyridine derivatives) possess highly correlated delocalized electrons at the origin of enhanced nonlinear optical effects. In POM like materials, the non-linear optical response can be accounted for by anharmonic distortion of the electron density distribution inside the molecule due to the intense electric field in a laser pulse.

It was proposed in a series of early papers by Robinson (1967), Flytzanis and Ducuing (1968) and Jha and Bloembergen (1968) that a set of simplifying models or approximations (Unsöld, 1927) applied far from resonance, would permit to connect susceptibilities to the various spatial moments of the charge distribution in the ground state. The model of Robinson (1967) offers the opportunity to calculate the polarizabilities and hyperpolarizabilities from the distribution of charges in the ground state which are at present time easily available from an X-ray experimental electronic density study. If the model of Robinson is working it is of primary interest to follow the discrepancies which may arise when we are replacing the isolated molecule picture by more realistic "in situ" counterpart as inferred from crystalline X-ray diffraction data. In this context our study is concerned with capabilities of an X-ray electronic density distribution to reveal crystalline effects on some well-defined optical nonlinear properties.

The first task is to evaluate the validity of the Robinson model on larger molecules using at the molecular level a semi-empirical quantum chemical approach implemented in MOPAC (MNDO, PM3) routine. From the semi-empirical atomic charges derived from a full Mulliken population analysis, the dipolar, quadrupolar and octupolar moments of POM molecule are inferred. Results of this so called "Point Charge" model allow for comparison of these different moments and corresponding derived polarizability tensor using the Robinson equations, with the equivalent NLO properties calculated from the same hamiltonian using the Finite Field method (Zyss 1979).

In the experimental approach values of the molecular quadrupole and octupole moments are estimated first from the multipolar coefficients of atomic density function as described by Hansen & Coppens (1978), and subsequently from the direct integration method using the two well documented space partitioning methods (Coppens, Moss & Hansen, 1980; Hirshfeld 1977).

The present study of the charge distribution and of one electron properties (namely the dipole, quadrupole and octupole moments) in a well documented nonlinear optical material, POM evidences interesting results with both confirmations and discrepancies as to earlier single or predictions. The connection between the dipole and quadrupole moments of charge distribution and the linear polarizability property through the Robinson model based on Unsöld approximation will be fully and critically discussed along this article.

Crystallographic Investigation

POM crystallizes in the non-centrosymmetric space group $P2_12_12_1$ with four molecules in the unit cell. The crystal data and results concerning the electron density distribution are published in previous paper (Hamzaoui, Baert & Zyss, 1996). The crystalline electronic density is described by the model presented in the introduction. The multipole populations at the convergence of the refinement are summarized in table I.

Table-I

The multipolar coefficients as described in the model of Hansen & Coppens. Pv correspond to l=0. d1, d2 and d3 to l =1 and Q1, Q2,Q3,Q4,Q5 are the quadrupolar coefficients corresponding to l=2 in the multipolar development. Some coefficients are set equal to zero because of atomic sites symetries.

Atoms	Pv	d1	d2	d3	Q1	Q2	Q3	Q4	Q5
N1	4.796	0.000	0.031	0.022	0.107	0.000	-0.023	-0.128	0.000
C2	4.337	0.000	-0.138	0.052	0.162	0.000	0.127	-0.124	0.000
C3	3.841	0.000	0.024	-0.003	0.095	0.000	-0.002	-0.165	0.000
C4	3.804	0.000	0.041	0.074	0.000	0.000	0.044	-0.140	0.000
C5	4.341	0.000	-0.046	-0.008	0.125	0.000	-0.028	-0.223	0.000
C6	4.273	0.000	-0.072	-0.057	0.118	0.000	0.094	-0.137	0.000
O7	6.221	0.000	0.007	-0.070	-0.157	0.000	-0.022	-0.060	0.000
C8	4.859	0.000	0.030	-0.121	-0.090	0.000	-0.001	0.011	0.000
N10	4.860	0.000	-0.001	0.076	0.039	0.000	-0.011	-0.148	0.000
O11	6.179	0.000	0.000	-0.033	-0.071	0.000	0.013	-0.110	0.000
O12	6.207	0.000	0.011	-0.017	-0.039	0.000	0.013	-0.093	0.000
HC2	0.660	0.000	0.000	0.072					
HC4	0.712	0.000	0.000	0.085					
HC5	0.593	0.000	0.000	0.077					
HC8	0.753	0.000	0.000	0.072					
HC8'	0.671	0.000	0.000	0.062					
HC8''	0.694	0.000	0.000	0.052					

MOLECULAR ELECTROSTATIC MOMENTS

From the knowledge of the density function one can derive some important physical properties of the molecules such as the surrounding electrostatic field gradient, and the different electrostatic moments of the charge distribution. A property associated to the average value of a quantum observable $\langle O \rangle$ is linked to the density function as given by the general equation (3), V is the molecular volume:

$$\langle O \rangle = \int_V \vec{O}(\vec{r}) \rho(\vec{r}) d\vec{r} \qquad (3)$$

If $\Delta\rho(\vec{r})$ rather than $\rho(\vec{r})$ is being considered the electrostatic moment due to the deformation density in the molecule and can be estimated. The property $\langle O \rangle$ can be approached by two different methods. In the first method, $\Delta\rho(\vec{r})$ was replaced by its corresponding expansion such as in the mutipolar model of Hansen & Coppens (1978) while in the second method it is being estimated directly using Fourier summation over all the structure factors. The second technique requires an accurate method for space partitioning.

The molecular dipole moment of POM bas been determined in previous paper (Hamzaoui & al. 1996) Such studies have clearly evidenced the electro donor character of the N-Oxide group in conjunction with the electro acceptor character of the nitro group. In general all the experimental methods agree on the same magnitude of about 1 Debye, which is about twice the ab initio value of 0.4 Debye found by Berthier, Defranceschi, Lazzeretti, Tsoucaris & Zenasi (1992). Values of the dipole moment used for the calculations are presented ine Table II.

Fig-1 : Orientation of the molecular dipole moment.

The positive sense from (charge excess) – to +(charge vacancy).

$\vec{\mu}_1$: From X-ray experiment

$\vec{\mu}_2$: From semi-empirical calculations

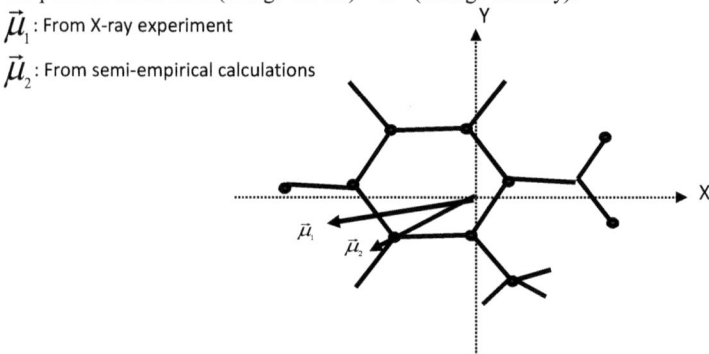

The components of the electrostatic quadrupole moment are obtained by substituting in equation (3) the operator $\hat{O}(r)$ by $\vec{r}_\alpha \wedge \vec{r}_\beta$. If in that equation the density function $\rho(\vec{r})$ is replaced by the multipolar expansion up to order $l = 1$, then the components of the quadrupole moment are given by :

$$Q_{\alpha\beta} = \sum \left[Q^i_{\alpha\beta} + \left(r_{i\beta} d_{i\alpha} + r_{i\alpha} d_{i\beta} \right) + r_{i\alpha} r_{i\beta} q_i \right] \qquad (4)$$

where $d_{i\alpha}$ and q_i represent respectively the a component of the dipole moment and the net charge of atom i at r_i . $Q^i_{\alpha\beta}$ are the atomic quadrupoles neglected here.

In the case of the direct integration method the development of equation (3) leads to:

$$Q_{\alpha\beta} = \frac{1}{V} \sum_{\vec{H}} \Delta F(\vec{H}) \left[\sum_i \left(Q^i_{\alpha\beta} + r_{i\beta} d_{i\alpha} + r_{i\alpha} d_{i\beta} + r_{i\alpha} r_{i\beta} q_i \right) \right] \qquad (5)$$

with

$$Q^i_{\alpha\beta} = \int_{ti} \left(r_\alpha - r_i \right) \left(r_\beta - r_i \right) e^{j2\pi \vec{H}(\vec{r} - \vec{r}_i)} d^3$$

The summation over $\vec{H}$ is performed over all structure factors and the indice t_i designates the integrable subunits. Evaluation of all molecular moments requires summations of the density and moments of each subunit which are being performed according to a space partitioning scheme. We used both discrete (Coppens, Moss & Hansen,1980) and fuzzy (Hirshfeld,1977) boundary space partitioning. The quadrupolar moment values are reported in the table-III- with the analogous components obtained from the point charge model using the net atomic charges derived by semi-empirical calculations (MOPAC 6). The most remarkable features when comparing experimental values with those derived from the free molecule stand-out in the Q_{xx}, Q_{yy}, Q_{xy} components. The experimental second moment component relative to a chosen molecular origin, ($Q_{xx} \approx -1.7$) taken as the average of the multipolar and direct integration values shows a weaker charge expansion than in the free molecule ($Q_{xx} \approx -4.8$) along the **X** direction, while the positive Q_{yy} 's indicate a similar contraction in the Y direction i.e. towards the molecular axis X (Fig. 1) for both the free molecule and the molecule in the crystal state. On the other hand the same electronic delocalization in the $(\vec{X}+\vec{Y})$ direction is being observed in the molecular plane for molecules in both states. Despite a higher dipole moment of the molecule embedded in the crystal we note smaller charge reorganization than in the free molecule, specially in the X direction.

Table-II

Components of the dipolar moment of the charge distribution (e Å) from the point charge model (the net charges are estimated by PM3 Hamiltonian) and comparison with the components derived from experimental electron density (Multipolar refinement and direct integration). The origin coincides with the center of mass of the molecule, and the cartesian system refered to the inertial axis of the molecule.

		μ_x (eÅ)	μ_y (eÅ)	μ_z (eÅ)	$\lVert\mu\rVert$ (Debye)
X-Ray Experiment	Multipolar Refinement	-0.18	-0.12	0.05	1.06
	Direct Integration	-0.22	-0.03	0.10	1.16
	Average Value	-0.20	-0.08	0.08	1.11
Point charge Model (MNDO)		-0.13	-0.12	0.04	0.88

Table-III

Components of the quadrupolar moment of the charge distribution (e Å^2) from the point charge model (the net charges are estimated by PM3 Hamiltonian) and compared to the components derived from experimental electron density. The origin coincides with the center of mass of the molecule, and the cartesian system refered to the inertial axis of the molecule.

	Semi-empirical Point charge Model	X-ray experiment			Values from α
		Multipolar Refinement	Direct Integration	Average values	
Q_{xx}	-4.820	-1.90	-1.48	-1.69	2.12
Q_{yy}	2.889	2.47	3.31	2.89	1.80
Q_{zz}	0.138	0.28	1.22	0.74	0.83
Q_{xy}	-0.662	-0.65	-0.54	-0.60	
Q_{xz}	-0.011	-0.09	-0.06	-0.08	
Q_{yz}	-0.266	-0.19	-0.18	-0.19	

MOLECULAR OPTICAL PROPERTIES

The energy of a molecule in an external field, $\vec{F}$, may be expanded in a power series as:

$$E(F)=E(0)-\mu_{i0}F_i -\frac{1}{2!}\alpha_{ij}F_I F_J -\frac{1}{3!}\beta_{ijk}F_i F_j F_k -\frac{1}{4!}\gamma_{ijkl}F_i F_j F_k F_l \ldots \qquad (8)$$

where summation over repeated indices is being assumed, μ_{i0} is the i component of the permanent dipole moment, α_{ij} is a tensor component of the linear polarizability and β_{ijk} is a component of the first order hyperpolarizability tensor. Since the nonlinear optical coefficients are functions of linear optical properties such as excitation energies and dipole matrix elements, the required electronic structure information can be conveniently obtained from spectroscopically based semiempirical electronic structure description. The combination of semiempirlcal molecular orbital methods and Finite Field perturbation technique (Zyss,1979; Kurtz, Stewart & Dicter, 1990), allow for the simultaneous calculations of all appropriate tensors components of the polarizability α and of the first and second hyperpolarizabilities (β and γ) for large organic molecules at the price of a moderate computational effort.

Values of α calculated by this method using the PM3 Hamiltonian are reported in table VI. The ratios between the diagonal components of the polarizability tensor (10^{-24} esu) and those of the quadrupolar moment (e $\mathring{A}^2$) (table III, colum 1) are of the order of 8 except for the smallest component α_{zz}. Those results seem to confirm roughly the Robinson's model based on the Unsöld approximation (1927) which allows to connect the susceptibility tensors of rang n to the multipolar charge distribution (in the fundamental state) of order 2^n. Thus, the linear polarizability is associated to the quadrupole moment, whereas the first order hyperpolarizability is linked to the octupolar moment of the charge distribution. The same type of relation between electrostatic moments and NLO properties has been evidenced also for the large NPP molecule (Fkyerat and al.1996).

This approximation was first applied to the model of linear molecular polarizability by Unsöld (1927) and generalized to nonlinear optics by Robinson (1967). Following Robinson (1967) one obtain for the linear polarizability:

$$\alpha_{ij}= \sum_{a\neq g} \frac{2e^2\Omega_{ag}}{\left(\Omega_{ag}^2-\hbar^2\omega^2\right)}\left\langle g\left|r_i\right|a\right\rangle\left\langle a\left|r_i\right|g\right\rangle \qquad (9)$$

where a and g refer to an excited and the ground state levels respectively. Ω_{ag} is the transition frequency between level a and the ground-state g, w is the frequency of the applied field. Variation of the transition dipole matrix elements $\left\langle g|r|a\right\rangle$ is generally very large compared to that of the energies Ω_{ag} in the denominator. One may therefore assume hat the energy terms are constant (Unsöld approximation). Calling Ω the mean value of he different Ω_{ag} energies leads to:

$$\alpha_{ij}=\frac{2e^2\Omega}{\left(\Omega^2-\hbar^2\omega^2\right)}\sum_{a\neq g}\left\langle g\left|r_i\right|a\right\rangle\left\langle a\left|r_j\right|g\right\rangle \qquad (10)$$

The linear polarizability is expressed directly in terms of the molecular quadrupole moments.

Following some straightforward matrix algebra and far from the resonance assuming that the frequency ω is negligible compared to the energy gap Ω, we obtain:

$$\alpha_{ij}=\frac{2e^2}{\Omega}\langle g|\hat{r}_i\hat{r}_j|g\rangle \quad (11)$$

Where $\hat{r}$ is defined as r-$\langle g|r|g\rangle$. The energy Ω can be obtained using the Thomas-Kuhn sum rule:

$$\sum_{a\neq g}\frac{2m\Omega_{ag}}{\hbar 2}|\langle a|r_i|g\rangle|^2=1 \quad (12)$$

Using the Unsöld approximation and notation given above, we obtain:

$$\frac{1}{\Omega}=\frac{2m}{\hbar^2}\langle g|\hat{r}^2|g\rangle \quad (13)$$

One can obtain for the three diagonal components (α_{xx}, α_{yy} and α_{zz}) similar equations as:

$$\frac{1}{\Omega_{xx}}=\frac{2m}{\hbar^2}\langle g|\hat{x}^2|g\rangle \quad (14)$$

Finally

$$\alpha_{xx}=\frac{4m}{\hbar^2}Q_{xx}^2 \quad (15)$$

The α_{xx} coefficient of the linear polarizability is then directly expressed in terms of the square component Q_{xx} ($Q_{xx}=\langle g|e\hat{x}^2|g\rangle$) of the second moment of the charge distribution. In the case of the off diagonal component, the α_{xy} components is obtained by considering the electric field F applied along the bisector of X and Y. This leads to replacement of $\hat{r}$ in Eq (10') by $\hat{r}=\frac{\hat{x}+\hat{y}}{\sqrt{2}}$, then:

$$\frac{1}{\Omega_{xy}}=\frac{m}{2\hbar^2}[\langle g|\hat{x}^2|g\rangle+\langle g|\hat{y}^2|g\rangle+2\langle g|\hat{x}\hat{y}|g\rangle] \quad (16)$$

by substitution in α_{ij} ($i=\hat{x}$ and $j=\hat{y}$) we get:

$$\alpha_{xy}=\frac{m}{\hbar^2}Q_{xy}\left(Q_{xx}+Q_{yy}+2Q_{xy}\right) \quad (17)$$

Table-VI

Values of the linear polarizability (10^{-24} e.s.u) from the finite field method (column I) and point charge model (column II) are compared to components components (column III) deduced from equations (11) and (12). (The profactor $\frac{4m}{\hbar^2}$ is equal to 7.6 . The quadrupole moment is in eÅ^2 and the linear polarizability in 10^{-24} e.s.u)

	I	II	II	IV	V
	Finite Field Method	Calculation Point charge Model	X-ray experiment		
			Multipolar refinement	Direct integration	Average Values
α_{xx}	36.12	176.56	27.43	16.64	22.03
α_{yy}	25.95	63.43	46.36	83.26	64.81
α_{zz}	5.55	0.14	0.59	11.31	5.95
α_{xy}	-0.865	8.18	-1.80	-1.47	-1.63
α_{xz}	0.149	0.19	0.61	0.03	0.32
α_{yz}	0.868	-2.52	-1.71	-2.85	-2.28
$\langle \alpha \rangle$	22.5	79.9	24.7	37.1	30.9

DISCUSSION:

Results based on these relations are summarized in table VI whereby the same components as infered from semi-empirical calculations have been included. Values in columns I and II expressed in the same units show that the "Point Charge Model' mimics to some relative extent the α values derived from the Finite Field method with coefficients ranging from 0.2 to 4.8 for the diagonal components. The average values of α are reported in the last row of table VI, the value 24.7 from the multipolar refinement is in good agreement with the corresponding one infered from the Finite Field method, however we observe an inversion in the magnitude of the α_{xx} and α_{yy} components calculated from both methods. This anomaly also reported by Fkyerat et al. (1996) in the case of NPP another prototypical paranitroaniline like NLO crystal (Zyss, Nicoud and Coquillay, 1984) could be 'a priori' related to crystal field effects.

However various possible sources of errors originating either from the models or from experimental interpretation or uncertainty could also be responsible for this discrepancy. An analysis of the origin of the discrepancies between the different results is to consider the diagonal components of the a tensor obtained from the Finite Field method as experimental observations (this hypothesis is justified by the nice agreement between calculated and measured values). The components of the quadrupolar moment Q_{ii} deduced from equation (15) and reported in Table III (last column), have the same order of magnitude as those derived from the X-ray experimental electronic density distribution with of course the inversion between the values of the components α_{xx} and α_{yy}. The discrepancy between the Q_{xx} component with respect to the similar one as derived from the X-ray experiment is related to the difference between the "isolated" and the "in situ" molecules where for this last one, strong perturbations due to the surrounding lattice (e.g. local fields and crystal field effects) are present. Such difference between "isolated" and "in-situ" molecules can be particularly enhanced along the main charge transfer axis x connecting the N-oxyde and nitro groups through the benzene ring, as a result of the larger polarizability in this direction. A depolarising field resulting from mutual polarization effects would be likely to be acting preferentially along the x axis and could lead to a reduction of the α_{xx} component. This observation would explain the strong value of Q_{xx} in the Point Charge model where the dipolar components have not been included.

However the inversion of the experimental magnitudes of α along the **X** and Y axis with respect to those infered from the Finite Field method is more troublesome and cannot received an explanation at present time. The experimental values have been of course calculated from the same set of data but with two quite different schemes (multipolar refinement and Fourier summation) and consequently the anomaly is quite independent of the method. At this stage concerning the linear polarizability as already evocated by Flytzanis (1975) the Unsöld approximation seems to give relatively good results for α.

CONCLUSION

The main goal of this work was first to check the validity of the Robinson-Unsöld model and secondly to test the capabilities of an X-ray electronic density distribution to derive linear polarizability through this model. The comparison of the components infered from the Finite Field method with the corresponding ones deduced from the Point Charge model show that the Unsöld approximation give to some relative extent good results for the α tensor, in agreement with the statement of Flytzanis (1975). The incorporation of the dipolar components in the calculation of the quadrupolar moment by the PC model should probably improve the results.

The results from the experimental electronic density study show a large under-estimate of the Q_{xx} value which leads through the Robinson model to a wrong determination of the polarizability. As these weak values have already been observed in the N-(4-nitrophenyl)-(L)-prolinol it is tempting to relate this anomaly to crystal effects, a kind of depolarising field particularly enhanced along the charge transfer axis. Despite the wrong order in the components of the polarizability and the calculation from the experimental results it is difficult at this stage to rely the discrepancies more to the model than the X-ray results. It is also puzzling to note that the crystal field effects are quite negligible regarding the determination of the NLO properties.

However, this study allows the calculation of the lowest-order macroscopic optical nonlinearity as the tensorial sum of molecular polarizabilities of the free molecule. The "oriented gas" description of the medium (Zyss and Tsoucaris, 1990 ; Zyss, Chemla, & Nicoud, 1981) applied to the obtained results will be presented and discussed in for coming paper.

References

1. Andreazza, P., Josse, D., Lefaucheux, F. Robert, M. C. & Zyss J. (1992). Phys. Rev. B45, 7640-7649.
2. Baert, F., Schweiss, P. Heger, G. & More, M. (1988) J. Mol. Struct. 178, 29-48.
3. Becker, P.J. & Coppens, P. (1974). Acta Cryst. A30, 129-147.
4. Berthier, G., Dfranceschi, M., Lazzeretti, P., Tsoucaris, G. & Zanasi, R. (1992). J. Mol. Struct. (Theochem), 254, 205-218.
5. Bessing, R. H. (1987). Cryst. Rev. 1, 3-58.
6. Boyd R.H. (1992). Nonlinear Optics, Academic Press Inc.
7. Brasselet, S. & Zyss, J. (1996). To be published.
8. Butcher, P. N. & Cotter, D. (1991). The Elements of Nonlinear Optics, Cambridge University Press.
9. Chemla, D. S. & Zyss, J. (1987). Nonlinear Optical Properties of Organic Molecules and Crystals, Vol. 1, Vol. 2, Academic Press, Inc.
10. Coppens, P., Moss, G. & Hansen N. K. (1980). Computing in Crystallography, Indian Academy of Sciences, p. 1601-1621.
11. Fkyrat, A., Guelzim, A., Baert, F. Zyss, J. & Perigaud, A. (1996) Phys. Rev. B53 16236-16246.
12. Flytzanis, Ch. & Ducuing, J. (1968). Phys. Rev. Vol. 178, 1218-1228.
13. Flytzanis, Ch. (1975). Quantum Electronics : A Tretise Vol. 1 Nonlinear optics, Part A edited by H. Rabin & C. L. Tang, p. 1-207. New York, San Francisco, London : Academic Press.
14. Hamzaoui, F., Baert, F. & Zyss, J. (1996). J. Mat. Chem. 6(7), 1123-1130.
15. Hansen, N. K. & Coppens, P. (1978). Acta Cryst. A34, 909-921.
16. Hirshfeld, F. K. (1977) Theoret. Chim. Acta, 129138.
17. Jha, S. S., & Bloembergen, N. (1968). Phys. Rev. Vol. 171, 891-898.
18. Kurtz, H. A., Stewart, J. J. P. & Dieter, K. M. (1990). J. Comp. Chem. 11(1), 82-87.
19. Kotler, Z., Hierle, R., Josse, D., Zyss, J. & Masse, R. (1992). Jocab. 9, 534-547.
20. Robinson, F.N.H. (1967). Bell System Technical Journal, 913-956
21. Robinson, F.N.H. (1968). J. of Physics, Proceeding of the Physical Society. 287-292
22. Unsold A. (1927) Z. Phys. 43, 563-574
23. Zyss, J. (1979a). J. Chem. Phys. 70, 3333-3340
24. Zyss, J. (1979b). J. Chem. Phys. 70, 3341-3349
25. Zyss, J; Chemla, D.S., & Nicoud, J.P. (1981). J. Chem Phys. 74, 4800-4811
26. Zyss, J., Nicoud, J.F. & Coquillay, M. (1984). J. Chem. Phys. 4160-4167
27. Zyss, J. & Tsoucaris, G. pp297-350 in Structure and Properties of Molecular Crystals, M. Pierrot (Ed.), Elsevier Amesterdam 1990.

Capability of X-ray Diffraction to the Determination of the Macroscopic Linear Susceptibility in a Crystalline Environment : the Case of 3-Methyl 4-Nitrropyridine N-oxide (POM)

Abdelkader Chouaih, Fodil Hamzaoui and Gérard Vergoten

Abstract

In a recent work we have reported the investigation of the applicability of ground state electron density inferred from X-ray diffraction to the determination of the linear polarizability of an "in situ" conjugated molecule labelled POM and corresponding to a nonlinear optical organic compound [J. Mol. Struc. Vol 397, Issues 1-3 (2004) 17-22]. In a such work and using the Unsöld approximation one can connect the susceptibility tensors of rang n to the multipolar charge distribution (in the fundamental state) of order 2^n. Thus, the linear polarizability is associated to the quadrupole moment, whereas be first order hyperpolarizability is linked to the octupolar moment of the charge distribution. Using a same approach as for the determination of the linear polarizability, we present here the estimated values of the tensor β derived from the X-ray experiment. However more surprising is the good agreement between the calculation of the lowest order macroscopic optical nonlinearity as tensorial sum of molecular polarizabilities of the free molecule and the macroscopic nonlinear measurements, the generalization of this remark would mean that the crystalline effects could be negligible in the estimation of the nonlinear optical properties.

However, it seems difficult to infer from the Unsöld approximation a faithful description of the molecular optical properties, whereas an estimation of the first order hyperpolarizability components of β seems out of reach.

Introduction

Considerable progress has been made in recent years towards understanding the nature of physical properties in term of the underling structures, In a previous lead article by Abrahams the main relations between properties of molecules embedded in the crystalline structures and corresponding physical properties have been extensively discussed [2]. For example the second harmonic generation (SHG) tensor coefficients of POM are two order of magnitude larger than in KDP a state of the art inorganic nonlinear crystal [3,4]. The linear macroscopic susceptibility of the crystal can be inferred from the molecular polarizability, based on oriented gas description model [5] and on the X-ray determination of the molecular structure and crystalline packing.

We have already reported the results on the linear polarizability obtained from x-ray diffraction study[1]. This connection is based on the application of a model due to Robinson [6,7] which allows to average the summation over excited states in the quantum expression of the polarizability to a single effective excited state [8], thereby reducing the expression of the polarizability to a compact formulation proportional to the average value of the electrostatic quadrupole tensor in the ground state. Assuming validity of this approach, the determination of the linear polarizability of the molecule should then be amenable to the knowledge of the ground state electronic charge distribution as inferred either from X-ray diffraction experiments, or from quantum calculations. We have demonstrated the applicability of this approach by using both X-ray diffraction and quantum chemical calculations and checking the consistency of both approaches with the Robinson quadrupolar model. Also, we have pointed-out to the limitations of this method along the polar directions which may be more sensitive to crystal field contributions.

Using the obtained results on the molecular linear polarizability and by application of the oriented gas model, it is possible to derive the index of refractions along the principal dielectric axis of the POM crystal which are along the three two-fold axis of the crystal [5]. An estimation of the first non linear hyperpolarizability will also be proposed and discussed along this article.

Crystallographic Investigation

The crystal data and results concerning the electron density distribution are published in previous papers [1,9,10]. Values and parameters of the Hansen and Coppens function [11] $\rho(\vec{r})$ at convergence are summarized in table-I. The different electrostatic moments of the charge distribution expressed by the property $\langle O \rangle$ can be directly derived from the knowledge of the density function $\rho(\vec{r})$ using the general equation given above :

$$\langle O \rangle = \int_V \hat{O}(\vec{r}) \rho(\vec{r}) d\vec{r} \qquad (1)$$

Where $\langle O \rangle$ is the property associated to the average value of a quantum observable linked to the density function, V is the molecular volume.

In a previous paper we have described the molecular components of the electrostatic dipole and quadrupole moments [1,10]. A such studies have clearly evidenced the connection between the molecular linear polarizability and the molecular electrostatic moment up the rank 2 of the charge distribution. In a similar fashion the electrostatic octupolar moment can be deduced by replacing the operator $\hat{O}$ by $\vec{r}_\alpha \vec{r}_\beta \vec{r}_\gamma$ in the *eq. (1)*, leading to:

15

$$O_{\alpha\beta\gamma}=\sum_i O^i_{\alpha\beta\gamma}+\sum_i\left(r_{i\alpha}Q_{i\beta\gamma}+r_{i\beta}Q_{i\alpha\gamma}+r_{i\gamma}Q_{i\alpha\beta}\right)+\sum_i\left(r_{i\beta}r_{i\gamma}d_{i\alpha}+r_{i\alpha}r_{i\gamma}d_{i\beta}+r_{i\alpha}r_{i\beta}d_{i\gamma}\right)+\sum_i r_{i\alpha}r_{i\beta}r_{i\gamma}q_i \quad (2)$$

The octupolar moment is given by the contribution of the atomic octupoles (first term), quadrupoles (second term), dipoles (third term) and net charges (last term) respectively. The different components were also estimated by both multipolar and direct methods. In the case of the multipolar model the first two terms were neglected (the multipolar development was stopped at l=1) while in the direct integration method all the terms have been taken into account. The corresponding expression after implementation of the direct integration method is:

$$O_{\alpha\beta\gamma}=\sum_{\vec{H}}\Delta F(\vec{H})\left[\sum_i O_{\alpha\beta\gamma}+\sum_i\left(r_{i\alpha}Q_{i\beta\gamma}+r_{i\beta}Q_{i\alpha\gamma}+r_{i\gamma}Q_{i\alpha\beta}\right)+\sum_i\left(r_{i\beta}r_{i\gamma}d_{i\alpha}+r_{i\alpha}r_{i\gamma}d_{i\beta}+r_{i\alpha}r_{i\beta}d_{i\gamma}\right)+\sum_i r_{i\alpha}r_{i\beta}r_{i\gamma}q_i\right] \quad (3)$$

where $O^i_{\alpha\beta\gamma}=\int_{t_i}\left(r_\alpha-r_i\right)\left(r_\beta-r_i\right)\left(r_\gamma-r_i\right)e^{j2\pi\vec{H}\left(\vec{r}-\vec{r}_i\right)}d^3r$ is the octupolar moment of the subunit t_i. The last integrals are well documented and expressed in terms of Bessel functions. The different electrostatic moments were estimated in the first method by the DSIGMA [11] program in which we have implemented the molecular higher order moments. For the direct integration method we have used the updated BARBIE program [12]. The results of the different calculations are summarized in table-II where the corresponding components estimated from the "Point Charge Model" have also been incorporated. One can note that the multipolar model and the direct integration methods lead to practically identical values. It is then confirmed that the atomic quadrupole and octupole moments do not contribute significantly to the molecular electrostatic moments. As for the Q_{xx} of the quadrupolar moment, we observe that the O_{xxx} component, is largely under-estimated with respect to the same component deduced from the "Point Charge Model". Some more comments about those anomalies are being deferred to the next section.

First order non linear hyperpolarizability

The energy of a molecule in an external field, $\vec{F}$, may be expanded in a power series as:

$$E(F)=E(0)-\mu_{i0}F_i-\frac{1}{2!}\alpha_{ij}F_I F_J-\frac{1}{3!}\beta_{ijk}F_i F_j F_k-\frac{1}{4!}\gamma_{ijkl}F_i F_j F_k F_l \ldots \quad (4)$$

where summation over repeated indices is being assumed, μ_{i0} is the i component of the permanent dipole moment, α_{ij} is a tensor component of the linear polarizability and β_{ijk} is a component of the first order hyperpolarizability tensor. Since the nonlinear optical coefficients are functions of linear optical properties such as excitation energies and dipole matrix elements, the required electronic structure information can be conveniently obtained from spectroscopically based semi-empirical electronic structure description. The combination of semi-empirical molecular orbital methods and Finite Field perturbation technique [13,14,15], allow for the simultaneous calculations of all appropriate tensors components of the polarizability α and of the first and second hyperpolarizabilities (β and γ) for large organic molecules at the price of a moderate computational effort.

Values of α and β calculated by this method using the PM3 Hamiltonian are reported in Table III and IV. The ratios between the diagonal components of the polarizability tensor (10^{-24} esu) and those of the quadrupolar moment (e $Å^2$) are of the order of 8 except for the smallest component α_{zz} while for the hyperpolarizability and the octupolar moment the coefficients range from 0.1 to 1.8. The same type of relation between electrostatic moments and NLO properties has been evidenced recently for the large NPP molecule [16].

Concerning the moments estimated from the X-ray electronic density study the Unsöld approximation gives relatively good results for α [17] but fails in the β estimate. The results reported here mimic very closely those published by Fekyerat et al. (1996) [16] on the N-(4-nitrophenyl)-(L)prolinol (NPP) another very efficient NLO compound.

By extension of the approach used for the linear polarizability, the expression of the first non linear hyperpolarizablty [18,19,20] is as follows :

$$\beta_{xyz} = \frac{e^3}{\hbar^2} \sum_P \sum_{a,b \neq g} \frac{\langle g|x|a\rangle \langle a|y|b\rangle \langle b|z|g\rangle}{(\omega_a - \omega - \omega_2)(\omega_b - \omega_2)} \quad (5)$$

where a and b are two different excited states. The summation over P express the simultaneous permutations over indices and frequencies, a complete expression of this last equation being given in the references just cited before. As developed in Appendix B the diagonal and non-diagonal components of the first non linear hyperpolarizability can be expressed in terms corresponding electrostatic octupolar and quadrupolar moments. The corresponding expressions are respectively :

$$\beta_{xxx} = 3\left(\frac{2m}{\hbar^2}\right)^2 Q_{xx}^2 O_{xxx} \quad (6)$$

$$\beta_{xyy} = \left(\frac{4m^2}{\hbar^2}\right)\left(Q_{xx}^2 + 2Q_{xx}Q_{yy}\right)O_{xxy} \quad (7)$$

Using these two last equations we are able to estimate the components of the first hyperpolarizability tensor, with results presented in table V, which provides also semi empirical values of the same components. The β tensor coefficients can be compared and analysed along similar lines as previously proposed in the linear case. The ratio between the β_{iii} infred from the Finite Field method and the components of the octupolar moment are ranging from 0.1 to 1.8

Following the same approach as in the linear case we consider the diagonal components of the β tensor calculated from Finite Field Method as experimental observations. Using Q_{ii} previously determinated we obtain the O_{iii} from eq. (6). The values thus deduced listed in column V of table IV fall off by one or two order of magnitude with respect to those found bye the point charge model and the schemes using the X-ray electronic density study, with no coherence at all between the magnitude and the signs in the components of the β tensor except for β_{xxx} the strongest components along the charge transfer axis (Table IV). As already mentioned in the linear case values from the point charge model (table IV, column II) are greater by two order of magnitude with respect to those deduced from the Finite Field method (table IV, column I), the most striking feature very much like in the case of the Alpha tensor, being an overestimate of β_{xxx} related certainly to a depolarizing field not taken into account in the point charge model.

The dispersion between the values of the F.F. method and those from the experimental measurements show that it is difficult to infer from the Unsold approximation a faithful description of the molecular optical properties, this is in agreement with the statement of Flytzanis (1975) [21] who predicted from theoretical basis an error by one to two order of magnitude in the estimation of β from the Unsold approximation. In the study of the N(4-Nitrophenyl)-(L)-prolinol (NPP) [16] the authors have also observed more than one order magnitude discrepancy between β values derived from F.F. method and P.C. model.

From the equation developed in appendix B the Robinson model shows how the charge asymmetry $\left\langle \hat{x}^3 \right\rangle$ and the charge expansion $\left\langle \hat{x}^2 \right\rangle$ intervene in the evaluation of β. Clearly as the system is getting larger, errors on the quadrupolar and octupolar molecular moments due to a crude estimation of $\Delta\rho$ or to defaults of the model, will increase like cube and the square of the size of the molecule, so it would not be surprising that the Robinson model failed for very large molecules and specially for the evaluation of β. The assumption included in the Unsold approximation that the same energies should be used o evaluate β, remain also a questionable one.

Macroscopic linear and non linear susceptibilities :

In a molecular crystal the "oriented gas" description [5] of the medium allows for straightforward connection of microscopic and macroscopic non lineartities, χ^1 are related to α, χ^2 to β and χ^3 to γ.

If, ij,k, are Cartesian indices related to molecular reference frame and I,J,K are Cartesian indices for the crystalline principal optical dielectric frame, we can write for χ^1 and χ^2 respectively :

$$\chi_{ij}^{(1)} = \frac{1}{V} F_I^{\omega} \sum_{i,j}^{n} \cos[I,i(s)] \cos[J,j(s))] \alpha_{ij}(s) \quad (8)$$

$$\chi_{ijk}^{(2)} = \frac{1}{V} F_I^{2\omega} F_J^{\omega} F_K^{\omega} \sum_{i,j,k}^{n} \cos[I,i(s)] \cos[J,j(s))] \cos[K,k(s))] \beta_{ijk}(s) \quad (9)$$

where $F_I^{2\omega}$, F_J^{ω}, F_K^{ω} are local field correction factors and V the volume of the unit cell. There are n molecules in the unit cell which are generated by the s elements of symmetry. The anisotropy correction of the local field is introduced through the refraction indices according the following equation :

$$F_I^{\omega} = \frac{n_I^2 + 2}{3} \quad (10)$$

The local field corrections according to equation (10) with the values of the refraction indices measured at 1.0632 μm provide three indices $n_b > n_a > n_c$ which could correspond to rather higher frequencies measurements (Table V). We also observe that the point charge model over estimate strongly the values of the refractive indices, however the proportionality between the different coefficients is nevertheless maintained.

As previously mentioned the Unsold approximation seems to provide good values for the linear polarization under the conditions of course that the moments of the charge distribution have been evaluated properly (Qxx is certainly largely overestimated in the PC model).

Results from moments inferred from the X-ray electronic charge distribution inverse the values of n_b and n_c due here to an underestimation of the component of the moments along X. These weak values are already been observed in the N(4-Nitrophenyl)-(L)-prolinol (NPP) another NLO charge transfer compound [6].

The hypothesis of depolarizing fields to explain this anomaly observed in the "in situ" molecule is particularly tempting, unfortunately the refractive indices calculated from these "in situ" values are in worst agreement with those inferred from the free molecule and the experimental measurements.

Concerning the macroscopic nonlinear succeptibility χ^2 the one dimensional model of Zyss and Oudar [22] gives for β_{xxx} a value of $(8.5\pm2)\times10^{-30}$ esu which agrees perfectly with 8.46 obtained from the semi empirical calculation. About the β's obtained from the experimental X-ray measurement using the Robinson model, the dispersion of the values shows that the model is clearly inadequate.

Conclusion

The comparison of the hyperpolarizablities components inferred from the Finite Field method with the corresponding ones deduced from the Point Charge model show that the Unsöld approximation gives results far less reliable in the estimation of β, in agreement with the statement of Flytzanis (1975) [21].

Concerning the macroscopic susceptibility and the determination of the refractive indices from the molecular polarizability inferred from the Finite Field method the agreement with the experimental measurements is largely satisfactory and appeared not fortuitous.

In view of the good results obtained for the macroscopic properties from semi-empirical calculation it seems very difficult through a model and from experimental electronic density distribution to obtain better values of the α tensor components and of the refractive indices. It is also puzzling to note that the crystal field effects are quite negligible regarding the determination of the NLO properties.

The main contribution of this work was to check the validity of the Robinson model (Unsöld approximation) and to follow the discrepancies which may appear when this model is applied both to a free molecule (Point Charge Model) and to the molecule in the crystal state. We have shown in this study that there is a reasonable agreement between electronic properties deduced from X-ray diffraction and those from semi-empirical calculations using the Point Charge Model. There is no doubt on some relations between the ground state molecular quadrupole and octupole and the polarizabilities of the molecule, however the Unsöld approximation in this study gives less reliable estimates of β than those claimed by O'Hare and al., Flytzanis and al. for smaller systems.

Of course it seems difficult to reach from the Usöld approximation a very faithful description of the molecular optical properties, an estimation of the first order hyperpolarizability components of β seems to be out of question.

19

Table-I : The coefficients obtained from the multipolar refinement.

Atoms	$l=0$	$l=1$			$l=2$				
	Pv	d1	d2	d3	Q1	Q2	Q3	Q4	Q5
N1	4.796	0	0.031	0.022	0.107	0	-0.023	-0.128	0
C2	4.337	0	-0.138	0.052	0.162	0	0.127	-0.124	0
C3	3.841	0	0.024	-0.003	0.095	0	-0.002	-0.165	0
C4	3.804	0	0.041	0.074	0.000	0	0.044	-0.140	0
C5	4.341	0	-0.046	-0.008	0.125	0	-0.028	-0.223	0
C6	4.273	0	-0.072	-0.057	0.118	0	0.094	-0.137	0
O7	6.221	0	0.007	-0.070	-0.157	0	-0.022	-0.060	0
C8	4.859	0	0.030	-0.121	-0.090	0	-0.001	0.011	0
N10	4.860	0	-0.001	0.076	0.039	0	-0.011	-0.148	0
O11	6.179	0	0.000	-0.033	-0.071	0	0.013	-0.110	0
O12	6.207	0	0.011	-0.017	-0.039	0	0.013	-0.093	0
HC2	0.660	0	0	0.072	0	0	0	0	0
HC4	0.712	0	0	0.085	0	0	0	0	0
HC5	0.593	0	0	0.077	0	0	0	0	0
HC8	0.753	0	0	0.072	0	0	0	0	0
HC8'	0.671	0	0	0.062	0	0	0	0	0
HC8"	0.694	0	0	0.052	0	0	0	0	0

Atoms	$l=3$						
	O1	O2	O3	O4	O5	O6	O7
N1	0.207	0	0.003	0.144	0	0	0.016
C2	0.280	0	-0.024	0.227	0	0	-0.015
C3	0.175	0	0.034	0.173	0	0	-0.021
C4	-0.237	0	-0.005	-0.126	0	0	-0.009
C5	0.261	0	-0.034	0.212	0	0	-0.017
C6	0.292	0	-0.009	0.267	0	0	0.006
O7	0.022	0	0.006	-0.012	0	0	0.010
C8	0.293	0	0.009	-0.037	0	0	-0.187
N10	0.261	0	0.004	0.166	0	0	0.021
O11	0.080	0	0.028	0.011	0	0	0.035
O12	0.071	0	0.047	-0.002	0	0	-0.002

Table-II : Components of the octupoalr moment of the charge distribution (e Å) from the point charge model (the net charges are estimated by PM3 Hamiltonian) are compared to the components derived from experimental electron density. The origin coincides with the center of mass of the molecule, and the cartesian system is being refered to the inertial axis of the molecule.

	Semi-empirical Point charge Model	X-ray experiment		
		Multipolar Refinement	Direct Integration	Values from β
O_{xxx}	4.641	0.35	-0.95	0.21
O_{xxy}	-1.484	-1.85	-1.52	
O_{xxz}	0.558	0.22	0.58	
O_{xyy}	-3.662	-3.66	-3.50	
O_{xyz}	-0.412	-0.17	-0.20	
O_{xzz}	0.104	0.28	0.29	
O_{yyy}	-2.163	-2.34	-1.90	0.009
O_{yyz}	0.054	-0.01	0.44	
O_{yzz}	-0.460	-0.78	-0.97	
O_{zzz}	0.082	0.15	0.34	0.001

Table-III : Values of the linear polarizability (10^{-24} e.s.u) from the finite field method (column I) and point charge model (column II) are compared to components (column III) deduced from equations (11) and (12). (The profactor $\dfrac{4m}{\hbar^2}$ is equal to 7.6 when the quadrupole moment is in $e\text{Å}^2$ and the linear polarizability in 10^{-24} e.s.u)

	I	II	II	IV	V
				X-ray experiment	
	Finite Field Method	Calculation P.C. Model	Multipolar refinement	Direct integration	Average Values
α_{xx}	36.12	176.56	27.43	16.64	22.03
α_{yy}	25.95	63.43	46.36	83.26	64.81
α_{zz}	5.55	0.14	0.59	11.31	5.95
α_{xy}	-0.865	8.18	-1.80	-1.47	-1.63
α_{xz}	0.149	0.19	0.61	0.03	0.32
α_{yz}	0.868	-2.52	-1.71	-2.85	-2.28
$\langle\alpha\rangle$	22.5	79.9	24.7	37.1	30.9

Table IV : Values of the molecular first hyperpolarizability (10^{-30} e.s.u.) from Finite Field method (column I) and Point Charge model (Comln II) are compared to components from equation (4) and (5) corresponding to the X-ray experiment (Column III). The profactor $\dfrac{4m^2}{\hbar^2}$ is equal to 3.01 when the electrostatic octupole moment is expressed in 10^{-30} e.s.u.

	I	II	III
	F.F. Method	P.C. Model	X-ray
β_{xxx}	8.46	973.63	-3.96
β_{xxy}	0.72	60.62	33.48
β_{xxz}	0.11	36.78	-0.37
β_{xyy}	-2.6	214.98	1.03
β_{xzz}	0.2	-0.41	0.08
β_{yyy}	0.26	-163.02	-158.42
β_{yyz}	-0.09	1.48	4.47
β_{yzz}	0.02	-1.13	-15.67
β_{zzz}	0.01	0.01	2.33

Table V : Values of refractive indices calculated from polarizability tensor components infred from the different methods (semi-empirical and experiment). In column IV the measured refractive indices at 1.0632 μm according to Zyss, Chemla and Nicoud [3].

	I	II	III	IV
	F.F. Method	P.C. Model	X-ray Experiment	Measured at 1.0632 μm
n_a	1.574	2.423	1.715	1.663
n_b	1.784	3.302	1.609	1.829
n_c	1.550	2.053	1.743	1.625

References:

[1] Hamzaououi F., Zanoun A and Veroten G. (2004) J. Mol. Struc. Vol 397, Issues 1-3, Pages 17-22

[2] Abrahams, S. C. (1994). Acta Cryst. A50, 658-685.

[3] Zyss, J; Chemla, D.S., & Nicoud, J.P. (1981). J. Chem Phys. 74, 4800-4811

[4]Zyss, J; Ledoux, I., Hierle, R.B., Raj, R.K. & Oudar, J.J. (1985). IEEE JQE 1286-1295

[5] Zyss, J. & Tsoucaris, G. pp297-350 in Structure and Properties of Molecular Crystals, M. Pierrot (Ed.), Elsevier Amesterdam 1990.

[6] Robinson, F.N.H. (1967). Bell System Technical Journal, 913-956

[7] Robinson F.N.H. (1968). J. of Physics, Proceeding of the Physical Society. 287-292

[8] Unsold A. (1927) Z. Phys. 43, 563-574

[9] Baert, F., Schweiss, P. Heger, G. & More, M. (1988) J. Mol. Struct. 178, 29-48.

[10] Hamzaoui, F., Baert, F. & Zyss, J. (1996). J. Mat. Chem. 6(7), 1123-1130.

[11] Hansen, N. K. & Coppens, P. (1978). Acta Cryst. A34, 909-921.

[12] Coppens, P., Moss, G. & Hansen N. K. (1980). Computing in Crystallography, Indian Academy of Sciences, p. 1601-1621.

[13] Zyss, J. (1979a). J. Chem. Phys. 70, 3333-3340

[14] Zyss, J. (1979b). J. Chem. Phys. 70, 3341-3349

[15] Kurtz, H. A., Stewart, J. J. P. & Dieter, K. M. (1990). J. Comp. Chem. 11(1), 82-87.

[16] Fkyrat, A., Guelzim, A., Baert, F. Zyss, J. & Perigaud, A. (1996) Phys. Rev. B53 16236-16246.

[17] Flytzanis, Ch. & Ducuing, J. (1968). Phys. Rev. Vol. 178, 1218-1228.

[18] Boyd R.H. (1992). Nonlinear Optics, Academic Press Inc.

[19] Chemla, D. S. & Zyss, J. (1987). Nonlinear Optical Properties of Organic Molecules and

[20] Butcher, P. N. & Cotter, D. (1991). The Elements of Nonlinear Optics, Cambridge University Press.

[21] Flytzanis, Ch. (1975). Quantum Electronics : A Tretise Vol. 1 Nonlinear optics, Part A edited by H. Rabin & C. L. Tang, p. 1-207. New York, San Francisco, London : Academic Press.

[22] Zyss, J. & Oudar, J.L. (1982). Phys. Rev. A26 2028-2048

SYNTHESIS AND MOLECULAR STRUCTURE INVESTIGATION BY DFT AND X-RAY DIFFRACTION OF ARNO

N. Benhalima, K. Toubal, A. Chouaih, G. Chita, S. Maggi, A. Djafri and F. Hamzaoui

Abstract

We report here the synthesis of (Z)-5-(4-nitrobenzyliden) -3-N(2- ethoxyphenyl)-2-thioxo-thiazolidin-4-one (ARNO) compound. The crystal structure has been determined by X-ray diffraction. The compound crystallizes in the triclinic system with space group $P\bar{1}$ and cell parameters: a = 9.1289(19), b = 9.3717(7), c = 12.136(3) Å, α=102.133(11)°, β=90.99(2)°, γ=117.165(9)°, V = 895.4(3) Å^3 and Z = 2. The structure has been refined to a final R = 0.05 for 2591 observed reflections. The refined structure was found to be significantly non planar. The molecule exhibits intermolecular hydrogen bond of type C–H...O and C–H...S. Ab initio calculations were also were performed at Hartree–Fock and density functional theory levels. The full HF and DFT geometry optimization was carried out using LANL2DZ, 6-31G* and B3LYP/6-31+G** basis sets. The optimized geometry of the title compound was found to be consistent structure determined by X-ray diffraction. The minimum energy of geometrical structure is obtained by using level HF/ LANL2DZ basis sets.

Keywords: Structure, X-ray diffraction, thiazolidin-4-one, ab initio calculations, ARNO.

Introduction

Research on new materials exhibiting nonlinear optical (NLO) behavior continues to be of primary interest for basic research as well as for industrial applications. The research on new materials with NLO properties for telecommunications and optoelectronics is directly related to the determination of their three-dimensional structure. Polymers represent a large family of interesting materials for nonlinear optics applications. In particular, compounds derived from thiazoles have recently received particular attention due to their NLO properties [1-3].

Density functional theory (DFT) is presently considered one of the most successful models in the world of computational chemistry since it yields accurate results for several physico-chemical properties, especially when hybrid DFT is used. The hybrid DFT functional offers reliable information for the excited state properties of small molecules [4], donor and acceptor systems [5], as well as metal complexes [6].

In this paper we present a structural study of the 3-(2-(1-hydroxycyclohexyl)-2-(4-methoxyphenyl) ethyl)-2-(4-methylphenyl)-thiazolidin-4-one (ARNO) compound by single-crystal X-ray diffraction to determine the most stable conformation in the crystalline state. To gain a better picture of the conformational profile of the given compound, we have also performed theoretical calculations using classical *ab initio* methods based on self-consistent field-molecular-orbital Hartree–Fock (HF) theory and Density Functional Theory (DFT) with the LANL2DZ, 6-31G* and 6-31+G** basis sets.

The results from X-ray diffraction have been compared to those obtained from *ab initio* DFT and HF calculations, finding a good agreement with the structure determined from the single-crystal measurements.

Experiment and computational methods

Synthesis

The title compound was prepared by reaction of N-arylrhodanine (0.01 molar), aldehyde (0.01 molar), 5 ml of acetic acid and sodium acetate (0.02 molar) in a 150 ml boiling flask. Then 2 ml of triethylamine are added to this mixture. The system is refluxed for 4 hours, forming a yellow solid. The crystals obtained are filtered and recrystallized in ethanol. Fig. 1 shows the chemical structure of the synthesized compound.

X-ray structure determination

A yellow prismatic crystal with approximate dimensions of 0.20 × 0.15 × 0.10 mm was selected for data collection. The X-ray diffraction data were collected on a Kappa CCD Nonius diffractometer. Reflection data were measured at 298 K using graphite monochromated MoKα radiation (λ = 0.71073Å). Intensities for 4080 reflections were measured with indices −11 < h < 11, −12 < k < 11, −15 < l < 15. The structure was determined by considering 2591 reflections with $I \geq 4\sigma(I)$. The structure was solved by direct methods using the SHELXS-97 [7]. A Fourier synthesis revealed the complete structure, which was refined by full-matrix least squares. All non-H atoms refined anisotropically. The positions of the H atoms bonded to C atoms were calculated. The H atoms were located from a difference Fourier map and included in the refinement with the isotropic temperature factor of the carrier atom. The final least-squares cycle using SHELXL-97 [8] gave R = 0.05 for the observed reflections with S = 0.95, $(\Delta\rho)_{min}$ = -0.425 eÅ^{-3}, $(\Delta\rho)_{max}$ = 0.219 eÅ^{-3}. An ORTEP [9] view of the molecular structure with the atomic numbering is shown in Fig 2. Atomic scattering factors for heavy atoms were taken from International Tables for X-ray Crystallography [10] while the factors for H were those of Stewart, Davidson & Simpson [11]. The details of crystal data and refinement are given in Table 1.

Table 1 Crystal data and structure refinement details

Compound	ARNO
Empirical formula	$C_{18} H_{14} O_4 N_2 S_2$
CCDC reference no.	805892
Formula weight	386.45
Crystal size (mm)	0.20 × 0.15 × 0.10
Temperature (K)	298(2)
Crystal system, space group	Triclinic, $P\bar{1}$
Unit cell dimensions	
a (Å)	9.1289(19)
b (Å)	9.3717(7)
c (Å)	12.136(3)
α (°)	102.133(11)
β (°)	90.99.(2)
γ (°)	117.165(9)
Wavelength (Å)	0.71073
Volume (Å³)	895.4(3)
Z, calculated density (mg/m³)	2/1.433
F(000)	400
θ range for data collection	5.01 – 27.50
Limiting indices	$-11 \leq h \leq 11, -12 \leq k \leq 11, -15 \leq l \leq 15$
Reflections collected/unique	4080/2591
Refinement method	full-matrix least-squares on F^2 data
Parameters	227
Goodness of fit on F^2	0.935
Final R indices [$F_0 > 4\sigma(F_0)$]	
R_1	0.0523
wR_2	0.1316
R indices (all data)	
R_1	0.0994
wR_2	0.1607

Computational method

For calculations involving hydrogen-bonding interaction systems it is very important to select an appropriate method, and carefully considering and evaluating its accuracy and speed of calculation. DFT methods are fast and can be used to compute mid-sized and even large molecular systems. In this work, full geometry optimization has been performed using the GAUSSIAN03 package [12] and the Gauss-View molecular visualization program [13], at the Becke3-parameter hybrid exchange functions and Lee-Young–Parr correlation functional (B3LYP) level [14–15] and HF theory [16], using the LANL2DZ, 6-31G* and 6-31+G** basis sets by the Berny method [17, 18].

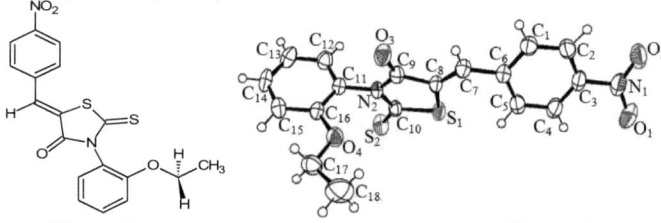

Figure 1	Figure 2

Fig. 1 Chemical structure of (Z)-5-(4-nitrobenzyliden) -3-N(2- ethoxyphenyl)-2-thioxo-thiazolidin-4-one (ARNO). **Fig. 2** General view of molecule with atomic numbering scheme (thermal ellipsoids drawn at 50% probability). H atoms are shown as small spheres of arbitrary radii.

Table 2 Bond distances for non-hydrogen atoms by X-ray and theoretical calculations (e.s.d.'s are given in parenthesis).

Bond distances (Å)	X-Ray	HF			B3LYP		
		LANL2DZ	6-31G*	6-31+G**	LANL2DZ	6-31G*	6-31+G**
S1 − C10	1.753(3)	1.809	1.760	1.760	1.846	1.785	1.782
S1 − C8	1.753(3)	1.807	1.761	1.760	1.820	1.761	1.761
S2 − C10	1.626(3)	1.667	1.629	1.629	1.678	1.640	1.640
O1 − N1	1.223(3)	1.238	1.193	1.194	1.280	1.231	1.232
O2 − N1	1.227(4)	1.239	1.193	1.194	1.280	1.231	1.232
O3 − C9	1.211(3)	1.216	1.186	1.187	1.244	1.213	1.215
O4 − C16	1.364(3)	1.368	1.340	1.339	1.388	1.356	1.357
O4 − C17	1.447(4)	1.447	1.410	1.412	1.471	1.432	1.435
N1 − C3	1.473(4)	1.461	1.458	1.461	1.474	1.470	1.471
N2 − C10	1.384(3)	1.371	1.362	1.364	1.388	1.378	1.380
N2 − C9	1.398(4)	1.403	1.394	1.393	1.424	1.412	1.410
N2 − C11	1.439(3)	1.438	1.432	1.433	1.447	1.437	1.437
C1 − C2	1.383(4)	1.388	1.380	1.382	1.398	1.387	1.389
C1 − C6	1.392(4)	1.404	1.395	1.396	1.423	1.413	1.414
C2 − C3	1.382(4)	1.390	1.383	1.383	1.406	1.394	1.395
C3 − C4	1.379(4)	1.390	1.382	1.382	1.406	1.394	1.395
C4 − C5	1.394(4)	1.389	1.382	1.383	1.399	1.389	1.390
C5 − C6	1.390(4)	1.404	1.394	1.395	1.422	1.412	1.413
C6 − C7	1.472(4)	1.472	1.474	1.474	1.462	1.455	1.456
C7 − C8	1.331(4)	1.334	1.327	1.328	1.361	1.353	1.354
C8 − C9	1.494(4)	1.487	1.491	1.493	1.492	1.492	1.493
C11 − C12	1.373(4)	1.383	1.376	1.377	1.398	1.389	1.389
C11 − C16	1.404(4)	1.397	1.394	1.395	1.414	1.408	1.408
C12 − C13	1.390(4)	1.395	1.386	1.388	1.407	1.395	1.397
C13 − C14	1.378(5)	1.392	1.381	1.383	1.406	1.393	1.395
C14 − C15	1.380(5)	1.396	1.388	1.389	1.407	1.397	1.398
C15 − C16	1.399(4)	1.393	1.386	1.388	1.409	1.399	1.401
C17 − C18	1.493(6)	1.519	1.514	1.514	1.524	1.518	1.518

Figure 3 A perspective view of the crystal packing in the unit cell.

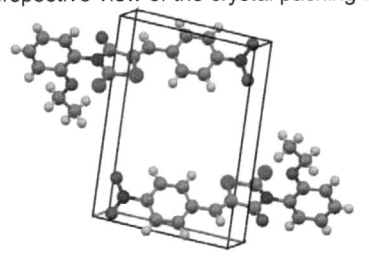

Table 3 Bond angles for non-hydrogen atoms by X-ray and theoretical calculations (e.s.d.'s are given in parenthesis).

Bond angles (°)	X-Ray	HF			B3LYP		
		LANL2DZ	6-31G*	6-31+G**	LANL2DZ	6-31G*	6- 31+G**
C10 − S1 − C8	93.12(13)	91.65	92.44	92.47	91.43	92.85	92.90
C16 − O4 − C17	118.0(2)	121.95	120.42	120.57	119.20	118.92	119.14
O1 − N1 − O2	124.2(3)	123.71	124.85	124.88	123.78	124.79	124.57
O1 − N1 − C3	118.3(3)	118.18	117.60	117.58	118.15	117.63	117.74
O2 − N1 − C3	117.5(3)	118.11	117.56	117.54	118.08	117.58	117.68
C10 − N2 − C9	117.1(2)	118.15	117.36	117.28	118.41	117.51	117.43
C10 − N2 − C11	122.1(2)	122.02	122.58	122.55	121.93	122.42	122.26
C9 − N2 − C11	120.8(2)	119.43	119.77	119.87	119.46	119.85	120.14
C2 − C1 − C6	121.4(3)	121.34	121.04	121.06	121.57	121.56	121.58
C3 − C2 − C1	118.6(3)	118.35	118.52	118.48	118.43	118.55	118.56
C2 − C3 − C4	122.1(3)	122.03	121.99	122.04	121.86	121.79	121.77
C2 − C3 − N1	119.6(3)	118.97	119.00	118.97	119.07	119.11	119.13
C4 − C3 − N1	118.3(3)	119.00	119.01	118.99	119.08	119.09	119.10
C3 − C4 − C5	118.0(3)	118.95	118.86	118.85	119.03	119.06	119.06
C6 − C5 − C4	121.6(3)	120.69	120.66	120.66	120.91	121.00	121.03
C5 − C6 − C1	118.2(3)	118.64	118.91	118.90	118.21	118.03	118.01
C5 − C6 − C7	123.1(3)	124.63	123.28	123.41	124.70	124.62	124.64
C1 − C6 − C7	118.7(3)	116.72	117.79	117.68	117.09	117.35	117.35
C8 − C7 − C6	128.8(3)	131.44	129.42	129.48	131.36	131.47	131.28
C7 − C8 − C9	121.8(3)	120.03	119.80	119.79	120.02	119.18	119.45
C7 − C8 − S1	128.5(2)	130.89	130.88	130.94	130.04	131.03	130.83
C9 − C8 − S1	109.47(19)	109.08	109.28	109.23	109.94	109.79	109.71
O3 − C9 − N2	122.8(3)	123.54	124.05	124.04	123.32	123.70	123.73
O3 − C9 − C8	127.2(3)	125.27	125.75	125.69	125.70	126.17	126.05
N2 − C9 − C8	110.0(2)	111.19	110.19	110.26	110.97	110.13	110.21
N2 − C10 − S2	128.0(2)	127.90	127.74	127.61	128.43	128.11	127.85
N2 − C10 − S1	110.13(19)	109.89	110.69	110.72	109.22	109.68	109.72
S2 − C10 − S1	121.91(17)	122.21	121.57	121.66	122.35	122.21	122.43
C12 − C11 − C16	121.4(3)	121.12	121.16	121.18	121.00	121.13	121.12
C12 − C11 − N2	120.9(2)	120.92	120.35	120.31	120.62	120.42	120.25
C16 − C11 − N2	117.7(2)	117.96	118.48	118.50	118.37	118.43	118.62
C11 − C12 − C13	119.5(3)	119.86	120.14	120.16	119.92	120.0	120.05
C14 − C13 − C12	119.7(3)	119.22	118.94	118.92	119.34	119.22	119.17
C13 − C14 − C15	121.4(3)	121.02	121.27	121.25	120.97	121.13	121.11
C14 − C15 − C16	119.6(3)	119.54	119.78	119.77	119.68	119.87	119.87
O4 − C16 − C15	125.7(3)	124.90	125.16	125.11	125.04	125.36	125.25
O4 − C16 − C11	115.9(2)	115.86	116.13	116.18	115.88	115.98	116.08
C15 − C16 − C11	118.4(3)	119.24	118.70	118.71	119.08	118.65	118.66
O4 − C17 − C18	106.9(3)	107.03	107.43	107.58	106.92	107.42	107.59

Results and Discussion

Description of the crystal structure

A general view of the molecule with atomic labeling (thermal ellipsoids are drawn at 50% probability) is shown in Figure 2. Figure 3 shows a perspective view of the crystal packing in the unit cell. Selected bond lengths, bond angles and torsion angles for all non − hydrogen atoms by X-ray diffraction are listed in Table 2, Table 3 and Table 4, together with the calculated parameters, respectively. The average values of bond distances and angles in the two benzene rings for both experimental and calculated are in good agreement with literature values. The three C−S

Table 4 Torsion angles involving non H-atoms by X-ray and theoretical calculations (e.s.d.'s are given in parenthesis).

Torsion angles (°)	X-Ray	HF			B3LYP		
		LANL2DZ	6-31G*	6-31+G**	LANL2DZ	6-31G*	6-31+G**
C6 − C1 − C2 − C3	−1.3(5)	−0.2	−0.92	−0.88	0.02	−0.20	−0.12
C1− C2 − C3 − C4	0.4(5)	−0.05	−0.26	−0.26	0.02	−0.04	0.01
C1 − C2 − C3 − N1	−179.3(3)	−179.9	−179.7	−179.7	−179.9	−179.9	−179.9
O1 − N1 − C3 − C2	−18.1(4)	−0.09	−0.38	−0.53	0.02	0.05	0.08
O2 − N1 − C3 − C2	161.5(3)	179.9	179.6	179.5	−179.99	−179.97	−179.9
O1 − N1 − C3 − C4	162.2(3)	−179.96	−179.8	180.0	−179.98	−179.8	−179.9
O2 − N1 − C3 − C4	−18.2(4)	0.02	0.18	0.00	0.01	0.17	0.12
C2 − C3 − C4 − C5	0.4(5)	0.16	0.68	0.68	−0.03	0.16	0.07
N1 − C3 − C4 − C5	−179.9(3)	−179.97	−179.9	−179.9	179.96	−179.98	−179.99
C3 − C4 − C5 − C6	−0.3(5)	−0.01	0.07	0.03	0.01	−0.04	−0.05
C4 − C5 − C6 − C1	−0.5(5)	−0.23	−1.21	−1.12	0.03	−0.19	−0.05
C4 − C5 − C6 − C7	−179.8(3)	−179.95	179.9	−179.98	−179.93	−179.99	−179.89
C2 − C1 − C6 − C5	1.3(5)	0.35	1.65	1.55	−0.04	0.31	0.14
C2 − C1 − C6 − C7	−179.3(3)	−179.9	−179.4	−179.5	179.92	−179.88	179.99
C5 − C6 − C7 − C8	21.0(5)	−4.77	−29.13	−28.14	0.17	−3.62	−1.95
C1 − C6 − C7 − C8	159.7(3)	175.5	151.98	152.98	−179.79	176.58	178.21
C6 − C7 − C8 − C9	−177.4(3)	−179.8	−178.7	−178.6	179.90	−179.96	−179.8
C6 − C7 − C8 − S1	−3.6(5)	−0.32	−1.38	−1.24	−0.01	−0.51	−0.40
C10 − S1 − C8 − C7	−170.7(3)	−179.8	−177.3	−177.5	179.76	−179.48	−179.5
C10 − S1 − C8 − C9	3.8(2)	−0.29	0.23	0.07	−0.16	0.01	−0.02
C10 − N2 − C9 − O3	178.2(3)	−178.95	−178.5	−178.7	−179.0	−178.34	−178.9
C11 − N2 − C9 − O3	−3.7 (4)	−5.95	−4.44	−4.74	−4.08	−3.57	−3.48
C10 − N2 − C9 − C8	−0.4(3)	1.87	2.36	2.15	1.61	2.14	1.61
C11 − N2 − C9 − C8	177.8(2)	174.9	176.4	176.1	176.5	176.90	177.0
C7 − C8 − C9 − O3	−6.2(5)	−0.29	−2.72	−2.43	0.02	−1.07	−0.80
S1 − C8 − C9 − O3	178.9(3)	−179.6	179.4	179.6	179.9	179.37	179.67
C7 − C8 − C9 − N2	172.3(3)	178.9	176.4	176.7	179.4	178.43	178.70
S1 − C8 − C9 − N2	−2.6(3)	−0.46	−1.41	−1.18	−0.70	−1.12	0.83
C9 − N2 − C10 − S2	−176.4(2)	178.3	178.3	178.4	178.4	178.21	178.8
C11 − N2 − C10 − S2	5.5(4)	5.46	4.42	4.64	3.58	3.59	3.48
C9 − N2 − C10 − S1	3.2(3)	−2.07	−2.19	−2.10	−1.70	−2.12	−1.62
C11 − N2 − C10 − S1	−174.1(19)	−174.9	−176.0	−175.8	−176.5	−176.74	−176.9
C8 − S1 − C10 − N2	−4.0(2)	1.27	1.04	1.08	0.99	1.13	0.88
C8 − S1 − C10 − S2	175.6(19)	−179.0	−179.4	−179.4	−179.1	−179.17	−179.47
C10 − N2 − C11 − C12	−82.7(3)	−89.9	−91.40	−91.56	−87.3	−91.15	−91.84
C9 − N2 − C11 − C12	99.2(3)	97.4	94.88	94.86	97.98	94.36	93.00
C10 − N2 − C11 − C16	98.4(3)	90.8	89.57	89.47	93.61	90.01	89.32
C9 − N2 − C11 − C16	−79.7(3)	−81.9	−84.1	−84.11	−81.1	−84.49	−85.84
C16 − C11 − C12 − C13	−0.5(5)	−0.33	−0.33	−0.39	−0.29	−0.31	−0.41
N2 − C11 − C12 − C13	−179.4(13)	−179.7	−179.3	−179.33	−179.4	−179.12	−179.2
C11 − C12 − C13 − C14	−1.3(5)	0.13	0.13	0.16	0.18	0.15	0.15
C12 − C13 − C14 − C15	1.2(5)	0.10	0.07	0.07	0.03	0.05	0.09
C13 − C14 − C15 − C16	0.6(5)	−0.13	−0.09	−0.09	−0.13	−0.09	−0.08
C17 − O4 − C16 − C15	−3.1(4)	1.95	2.55	1.68	−0.50	0.93	0.02
C17 − O4 − C16 − C11	177.4(3)	−178.3	−177.6	−178.4	179.5	−179.11	−179.96
C14 − C15 − C16 − O4	178.1(3)	179.7	179.7	179.79	−179.96	179.89	179.85
C14 − C15 − C16 − C11	−2.4(4)	−0.06	−0.10	−0.12	0.02	−0.06	−0.17
C12 − C11 − C16 − O4	−178.1(3)	−179.5	−179.5	−179.6	−179.8	−179.70	−179.60
N2 − C11 − C16 − O4	0.8(4)	−0.15	−0.53	−0.60	−0.71	−0.86	−0.77
C12 − C11 − C16 − C15	2.4(4)	0.29	0.3	0.36	0.19	0.26	0.42
N2 − C11 − C16 − C15	178.7(2)	179.7	179.3	179.3	179.3	179.10	179.25
C16 − O4 − C17 − C18	−176.1(3)	179.01	178.5	179.0	−179.8	179.18	179.92

distances, S1−C8, S1−C10 and S2=C10 [1.753(3), 1.753(3) and 1.626(3) Å], respectively in the thiazole ring have values intermediate between those reported for $C(Sp3)−S$ single [1.81 Å] and double [1.61 Å] bonds [10]. The mean value of bond angles in thiazole ring is 107.96 (2)°. The crystal structure exhibits intermolecular interaction of the type C−H...O and C−H...S in which C atoms (C2, C4, C5, C7, C13 and C14) act as donors and O (O1, O2 and O3) and S1 atoms acts as acceptors. In the crystalline state, these intermolecular interactions stabilize the crystal structure. The geometry of the hydrogen-bonded interactions is listed in Table 5. Fig. 4 shows some hydrogen bonds in the crystal. All bond angles C−C−C, C−N−C and C−C−N,... are close to 120°, indicating that the π electrons in the ARNO molecule are delocalized.

Crystallographic data (excluding structure factors) for the structure reported in this article have been deposited with the Cambridge Crystallographic Data Centre as supplementary publication number CCDC 805892 [19].

Fig. 4 View of the two H–bonds C7 − H7...O3 in the ARNO crystal.

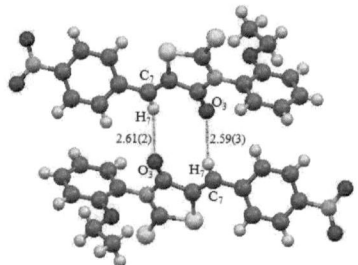

Table 5 Geometry of the C − H...O and C − H...S hydrogen bonds in ARNO crystal by X-ray diffraction.

D − H...A	D − H (Å)	H...A (Å)	D...A (Å)	D − H...A (°)
C2 − H2...O1	0.93 (3)	2.49 (3)	2.761 (5)	96.5 (3)
C4 − H4...O2	0.93 (4)	2.45 (2)	2.724 (4)	96.9 (3)
C5 − H5...S1	0.93 (3)	2.55 (1)	3.198 (3)	127.0 (2)
C7 − H7...O3	0.93 (4)	2.61 (2)	2.936 (4)	101.1 (2)
C4 − H4...O3^{i}	0.93 (4)	2.93 (3)	3.278 (6)	103.4 (3)
C7 − H7...O3^{ii}	0.93 (4)	2.59 (3)	3.267 (4)	129.5 (2)
C13 − H13...O2^{iii}	0.93 (3)	2.80 (3)	3.385 (5)	121.9 (3)
C14 − H14...O2^{iii}	0.93 (4)	2.76 (3)	3.373 (5)	124.0 (2)

Symmetry code: *(i) x−1,+y,+z, (ii) −x+2, −y, −z, (iii) x+2,+y+1,+z*

Geometry optimization
The ground state geometries were optimized by the Hartree Fock and DFT levels of theory, using LANL2DZ, 6-31G(d) and 6-31+G(d,p) basis sets. The selected optimized bonds lengths, bond angles and torsion angles are given in Tables 2, 3 and 4, as we can see there is a good agreement between the calculated and the experimental values. We also checked the effect of basis sets on the calculations. The largest deviation between X-ray data and theoretical calculations at the HF/LANL2DZ level is the S1−C10 distance, around 0.06 Å, and the C16−O4−C17 angle, which is larger than 3°. The B3LYP/LANL2DZ results deviate in the range from 0.001 to 0.9 Å for bond lengths, and from 0.02° to 2.56° (C8−C7−C6) for bond angles. The difference between the experimental and calculated bond lengths calculated at the HF level with 6-31G(d) basis set does not exceed 0.037Å (O4−C17), whereas in the case of B3LYP with same basis set, the largest difference between the observed and the calculated values is about 0.03 Å. The bond angles for HF/631G(d) calculations are very close to the experimental values (Table 3), and the maximum difference is about 2.42°. For DFT with 6-31G(d) basis the bond angle difference does not exceed 2.67°. The HF/6-31+G(d,p) and B3LYP/6-31+G(d,p) results deviate in the range from 0.001 to 0.035Å (O4−C17) and 0.001 to 0.029Å for the bond lengths, and from 0.04° to 2.57° (C16−O4−C17) and 0.04° to 2.48° (C8−C7−C6) for the bond angles, respectively.

Fig. 5 Theoretical crystal structure of ARNO with B3LYP/6-31G(d) level.

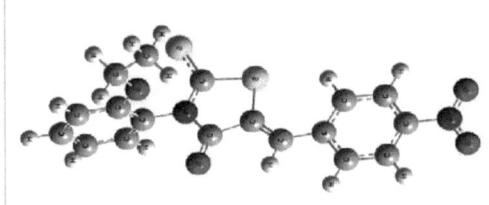

In summary, the optimized bond lengths and bond angles obtained using the DFT method are in good agreement with the corresponding X-ray structural parameters. It is worth noting that some of the optimized torsion angles have slightly different values from the corresponding experimental ones, due to the fact that the theoretical calculations consider only isolated molecules in the gaseous phase while the experimental results refer to molecules in the crystal environment.

Conclusions

In this study, we have synthesized the (Z)-5-(4-nitrobenzylīden) -3-N(2-ethoxyphenyl)-2-thioxo-thiazolidin-4-one (ARNO) compound and its crystal structure was determined by X-ray diffraction. This compound belongs to the centrosymmetric space group $P\bar{1}$. From the crystal structure, this compound seems to be potentially useful for non-linear optical applications. The results from X-ray diffraction were assessed by DFT and HF ab initio calculations using three different basis set LANL2DZ, 6-31G* and B3LYP/6-31+G**. The two ab initio computational methods gave very similar results, which are very close to those of X-ray data.

In a forthcoming paper, we will present a spectroscopic study and other theoretical calculations on the same compound in order to evaluate the main physico-chemical properties, such as the atomic charge distribution and the dipole moment, necessary to assess the efficiency and applicability of the title compound in the nonlinear optics field.

References

[1] Smokal V, Derkowska B, Czaplicki R, Krupka O, Kolendo A, Sahraoui B (2009) Optical Materials 31:554-557
[2] Smokal V, Kolendo A, Derkowska B, Czaplicki R, Krupka O, Sahraoui B (2008) *Molecular Crystals and Liquid Crystals* 485:1011-1018
[3] Watanabe T, Amano M, Tomaru S (1994) Jpn. J. Appl. Phys. 33:L1683-L1685
[4] Adamo C, Barone V (1998) J. Chem. Phys. 108 :664
[5] Jamorski-Jödicke C, Lüihi HP (2002) J. Chem. Phys. 117:4146
[6] Cavillot V, Champagne B (2002) Chem. Phys. lett. 354:499
[7] Sheldrick GM (1997) SHELXS97. Program for crystal structure determination, University of Göttingen. Germany
[8] Sheldrick GM (1997) SHELXL97. Program for crystal structure refinement, University of Göttingen. Germany
[9] Farrugia LJ (1997) J. Appl. Crystallogr. 30:565
[10] Wilson AJC (1995) International Tables for Crystallography, Volume C. Kluwer Academic Publishers, Dordrecht, The Netherlands
[11] Stewart RF, Davidson ER, Simpson WT (1965) J. Chem. Phys. 42:3175-3187
[12] Frisch MJ, Trucks GW, Schlegel HB, Scuseria GE, Robb MA, Cheeseman JR, Montgomery JA Jr, Vreven T, Kudin KN, Burant JC, Millam JM, Iyengar SS, Tomasi J, Barone V, Mennucci B, Cossi M, Scalmani G, Rega N, Petersson GA, Nakatsuji H, Hada

M, Ehara M, Toyota K, Fukuda R, Hasegawa J, Ishida M, Nakajima T, Honda Y, Kitao O, Nakai H, Klene M, Li X, Knox JE, Hratchian HP, Cross JB, Adamo C, Jaramillo J, Gomperts R, Stratmann RE, Yazyev O, Austin AJ, Cammi R, Pomelli C, Ochterski JW, Ayala PY, Morokuma K, Voth GA, Salvador P, Dannenberg JJ, Zakrzewski VG, Dapprich S, Daniels AD, Strain MC, Farkas O, Malick DK, Rabuck AD, Raghavachari K, Foresman JB, Ortiz JV, Cui Q, Baboul AG, Clifford S, Cioslowski J, Stefanov BB, Liu G, Liashenko A, Piskorz P, Komaromi I, Martin RL, Fox DJ, Keith T, Al-Laham MA, Peng CY, Nanayakkara A, Challacombe M, Gill PMW, Johnson B, Chen W, Wong MW, Gonzalez C, Pople JA (2003) Gaussian 03, Revision B03, Gaussian Inc., Pittsburgh PA

[13] Frisch AE, Nielsen AB, Holder AJ (2003) Gaussview, Gaussian Inc., Carnegie Office Park, Building 6, Pittsburg, PA 15106, USA
[14] Becke AD (1997) J. Chem. Phys. 107:8554
[15] Rauhut G, Pulay P (1995) J. Phys. Chem. 99:3093-3100
[16] Cohen HD, Roothaan CC (1965) J. Chem. Phys. 43:S34-S39
[17] Fletcher R, Powell MJD (1963) Comput. J. 6:163
[18] Bader RF (1990) Atoms in Molecules. A Quantum Theory. Clarendon Press, Oxford
[19] Copies of the data can be obtained free of charge on application to CCDC, 12 Union Road, Cambridge CB2 1EZ, UK. Fax: +44 1223 336033; e-mail: *deposit@ccdc.cam.ac.uk*

ELECTRON CHARGE DENSITY DISTRIBUTION FROM X-RAY DIFFRACTION STUDY OF THE M-NITROPHENOL COMPOUND IN THE MONOCLINIC FORM

Fodil Hamzaoui , Mokhtaria Drissi , Abdelkader Chouaih , Philippe Lagant and Gérard Vergoten

Abstract:
At room temperature, the m-Nitrophenol (m-NPH) appears in two polymorphic structures: orthorhombic and monoclinic forms. In the present work, we shall focus on the monoclinic form of this compound which has a centrosymmetric structure with the space group $P2_1/n$. The molecular dipole moment has been estimated experimentally. High resolution single crystal diffraction experiment was performed at low temperature with $MoK\alpha$ radiation. The crystal structure was refined using the multipolar model of Hansen and Coppens (1978). The molecular electron charge density distribution is described accurately. The study reveals the nature of inter-molecular interactions including charge transfer and hydrogen bonds. In this crystal, hydrogen bonds of moderate strength occur between the hydroxyl group and the O atom in the nitro one.
Keywords: Electron charge density, M-Nitrophenol, XD program, nonlinear optical compound (NLO)

1. Introduction

It is known that m-Nitrophenol crystallizes into two polymorphic forms (monoclinic P21/n and orthorhombic P212121), with four molecules in the unit cell [1-3]. The present work focuses on the structural and electronic charge density study of the molecule of the title compound in the monoclinic phase with the space group P21/n. The structure was first described by Panadares et al. (1975) [4].

To achieve our work we have used the XD package software [5], for a non-spherical atom multipole refinement which has been developed by Hansen and Coppens (1978) [6]. One major component of this package is the program for the least square fitting of multipole model to the experimental data [7-8].

The thermal motion and the structure analysis of the molecule have been performed. The electron density maps have been provided at different sections of the molecule. We also have made available the electron charge distribution around the hydrogen bond and calculated the molecular dipole moment.

Accurate results on the structure and electron charge density distribution of the m-nitrophenol have been exposed in details along this article which highlights the adequacy of the multipolar model and theoretical calculations.

2. *Experimental* investigation

As for the previous compounds, the crystallographic investigation have been carried out using the Hansen-Coppens multipole formalism [6] described in the introduction. To reduce the number of the parameter refinements the locol symmetries shown in figure 1 are assumed.

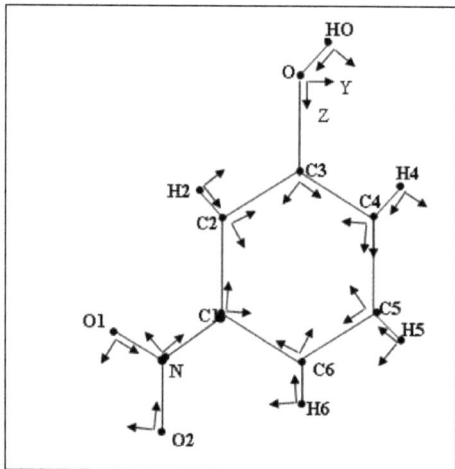

Figure 1. Labeling of the atoms and definition of local orthogonal reference axes for the atom-centered multipolar functions

3. Results and Discussion

3.1. Structural and Thermal vibration analysis

The general features of the structure have been described previously [3]. The main characteristic of this structure is that the four molecules in the unit cell are engaged in four chains formed by infinite chains of hydrogen-bonded coplanar molecules. The H atoms were placed by setting the bond distances C—H and O—H at 1.105 Å and 1.029 Å respectively. Any position error of the hydrogen atom will bring about errors strongly correlated with the dipolar population of hydrogens.

The thermal motion analysis of m-NPH has been performed using the THMA11 program [14]. The rigid-body motion is described by three tensors T, L and S taking into account for translation, libration and the correlation between translation and libration, of the rigid group respectively. These tensors are obtained by a least-squares fit refinement using the observed atomic thermal motion parameters obtained by the refinement. This test indicated that the differences between the mean square displacements of atoms (MSDAs) along interatomic directions have magnitudes $\Delta \leq 10.10^{-4}\text{Å}^2$ for the bonded pairs of the non-H atoms [15]. The MSDAs Δ_{AB} in the AB direction for all pairs of atoms in the molecule have been calculated and reported in Table 1.

Table 1. Matrix for differences in MSDA's (mean square displacements of atoms) [Values listed are 10^4 MSDA's for column atom minus that for row atom, underlined values correspond to chemical bonds].

	O	O2	O1	N	C6	C5	C4	C3	C2
C1	-5	5	2	14	5	17	16	10	-4
C2	-5	27	-9	-1	7	7	8	11	
C3	-10	16	-10	-7	8	5	1		
C4	-19	5	-1	-2	4	4			
C5	-19	-13	0	-7	-4				
C6	-18	-17	5	-3					
N		-4	-15	-7					
O1		2	5						
O2	-40								

In the general treatment of the molecular thermal motion in terms of rigid-body TLS, the calculated anisotropic thermal parameters are given in the Trueblood notation [14] as such:

$$U_{ij} = T_{ij} + G_{ijkl}L_{kl} + H_{ijkl}S_{kl} + D^2\Omega^2 n_i n_j$$

(3)

Where G and H are geometrical parameters and S is an asymmetric tensor needed to account for the average quadratic correlation of T and L. The last term corresponds to any additional intra-libration (Ω) around a chosen axis. The rigid-body fit suggests an independent liberation axis around the C1—N bond. The thermal motion of the H atoms is considered to consist of two contributions. The first is due to rigid molecular motion and the second is from the C—H vibrations [16-18].

The ellipsoids of the different atoms representing their thermal motion described above are shown with an ORTEPIII diagram [19] in Figure 2.

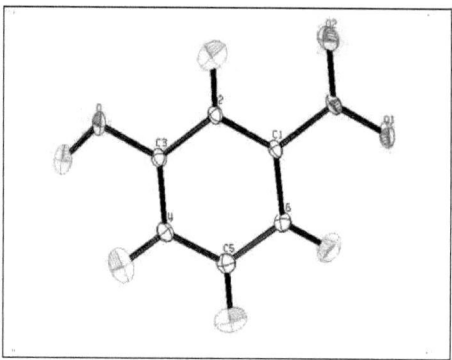

Figure 2. *ORTEPIII* (Johnson, 1996) diagram of the m.Nitrophenol molecule

3.2. Electron-density maps

The aspherical atom model used in multipole refinement gives structure factor phases closer to the true phases for crystals than the spherical or independent atom model does. This enables the mapping of the electron density by Fourier synthesis in various ways using the program XDGRAPH implemented in the XD program package [5].

The experimental density deformation map is shown in Figure 3, from which we can notice the absence of the density on the atomic sites and the appearance of all the bond density peaks and also the localization of the oxygen lone pairs of the nitro group. This map confirms the high quality of the data sets and the efficiency of the formalism used for the data processing as proposed by Blessing [20]. This visualization is obtained using the calculated multipole phases with the observed structure factors F_{obs} (h):

$$\delta\rho^{exp}(r) = \frac{1}{V}\sum_{h}\left[\left|F_{Obs}(h)\right|.e^{i\phi_{mul}} - \left|F_{sph}(h)\right|.e^{i\phi_{sph}}\right]e^{-2\pi i h.r} \qquad (4)$$

$Fsph$ (**h**) is computed with atomic positions and thermal parameters obtained from the multipole refinement. The electron density deformation obtained from the last refinement is the dynamic model map. This map is obtained from the calculated multipole structure factors, *i.e.* the Fourier coefficients are the difference of two values of F_c:

$$\delta\rho^{dyn}(r) = \frac{1}{V}\sum_{h}\left[\left|F_{mul}(h)\right|.e^{i\phi_{mul}} - \left|F_{sph}(h)\right|.e^{i\phi_{sph}}\right]e^{-2\pi i h.r} \qquad (5)$$

Temperature factors are included in F_{mul} and F_{sph}. This density distribution is free of experimental noise. Figure 4 shows this electron density deformation, where one can easily observe the obvious increase of the density peaks and the good localization of the oxygen atoms lone pairs O1 and O2. The presented maps are given in the benzene ring section and the contour map is 0.05 e$Å^{-3}$.

A residual density map in the molecular plane obtained in the final cycle of refinement (see Figure 5) shows the adequacy of the multipolar model to describe the electron experimental density of the molecule. The absence of the quasi-totality of the density peak again confirms the high quality of the recorded data and the precision of the used equipment. On the other hand, the multipole expansion model appears to be very efficient for describing the electron density distribution in structure [21-22].

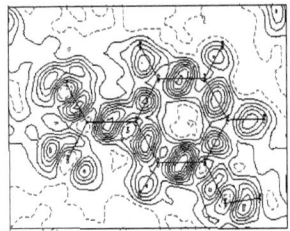

Figure 3. Experimental density map from high-order refinement: $\rho_{\exp} = \rho_o - \rho_{sph}$. ρ_o is the observed electron density and ρ_{sph} is the calculated electron density using the atomic parameters obtained from the high-order refinement. Contour map is 0.05 eÅ^{-3}

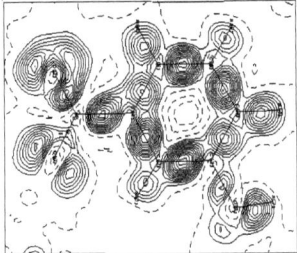

Figure 4. Dynamic density map: $\rho_{dyn} = \rho_{mult} - \rho_{sph}$, ρ_{mult} is the calculated electron density using the multipolar model and ρ_{sph} is the calculated electron density using the atomic parameters obtained from the high-order refinement. Contour map is 0.05 eÅ^{-3}

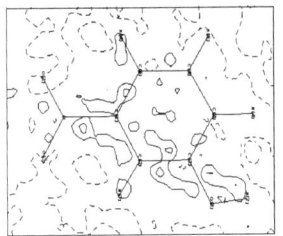

Figure 5. Residual density map: $\rho_{res} = \rho_o - \rho_{mult}$. Contour map is 0.05 eÅ^{-3}

3.3. Molecular dipole moment

The molecular dipole moment was calculated from the multipolar population parameters (Table 2), following the procedure described by Hansen and Coppens (1978) [6]. Values of the multipole parameters are summarized in Table 3. The value of the dipolar moment reaches 5.81 Debye, see Table 4. Its orientation in the molecule is shown in Figure 6. The method is in accordance with the evaluation of the positive sign of the net charges on the H atoms and the negative sign of the net charges on the O atoms.

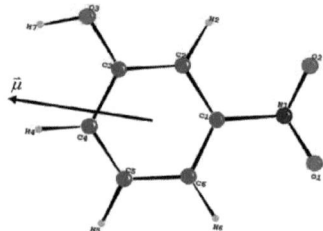

Figure 6. Molecular dipole moment calculated by multipolar model. The origin is at the center of mass of the molecule

Table 2. Net atomic charge in m-Nitrophenol

ATOM	P_v	q
C1	4.153	-0.153
C2	4.270	-0.270
C3	3.971	0.028
C4	4.395	-0.395
C5	4.462	-0.462
C6	4.268	-0.268
N	4.353	0.646
O1	6.233	-0.233
O2	6.218	-0.218
O	6.290	-0.290
H2	0.681	0.318
H4	0.783	0.216
H5	0.749	0.250
H6	0.711	0.288
HO	0.671	0.328

Table 3. The coefficients obtained from the multipolar refinement

Table 4. Magnitude of the molecular dipole moment

μ_x	μ_y	μ_z	μ (Debye)
-0.032	-0.032	-0.636	5.81 (20)

3.4. Hydrogen bonds

The crystal structure of the mNPH rests on chains of molecule joined by hydrogen bonds, the almost linear hydrogen bonding links by translation of equivalent molecules along the c crystallographic axis through the OH and NO_2 groups (see Figure 7). Hydrogen atoms are positioned to give 1.105 and 1.029 Å C—H and O—H bonds lengths, respectively. The O—H----O distance is a little shorter in the orthorhombic crystal (2.88Å), than in the monoclinic structure (2.94Å). On the other hand the O—H----O angles differ slightly: 168° and 178° in both polymorphic structures respectively. There are nine intermolecular interactions which are possible in the hydrogen bond network as shown in table 5.

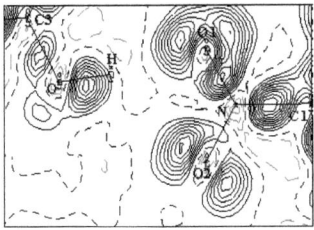

Figure 7. Deformation density map in the plan of the hydrogen bond; Contour map is 0.05 eÅ^{-3}

Table 5. Various possible hydrogen bonds

D —H ...A	D —H...A (°)	D —H (Å)	D...A (Å)	H....A (Å)
C2—H2.....O2 (a)	95.26 (8)	1.105 (2)	2.717 (3)	2.382 (3)
C6—H6.....O1 (a)	91.79 (9)	1.093 (2)	2.727 (3)	2.464 (2)
C4—H4.....O1 (b)	129.70 (9)	1.094 (2)	3.386 (4)	2.581 (3)
O —HO.....N (b)	155.47 (8)	1.029 (1)	3.495 (4)	2.532 (3)
O —HO.....O1 (b)	178.03 (8)	1.029 (1)	2.908 (3)	1.880 (2)
O —HO......O2 (b)	127.49 (8)	1.029 (1)	3.315 (4)	2.586 (3)
C5—H5.....O1 (c)	124.62 (9)	1.109 (2)	3.693 (5)	2.949 (4)
C5—H5.....O2 (d)	140.67 (9)	1.109 (9)	3.338 (3)	2.405 (2)
C5—H5.....O (e)	131.87 (9)	1.109 (2)	3.471 (3)	2.631 (2)

Symmetry operations
(a) *x, y, z*   (b) *x, +y, +z-1*   (c) *-x+2, -y+1, -z+1*
(d) *x+1/2, -y+1/2, +z-1/2*   (e) *x+1/2, -y+1/2, +z+1/2*

4. Experimental Section

The crystallographic data were obtained at 122K on a CAD4 diffractometer using the graphite–monochromated MoKα radiation. The crystal of high quality was cooled by using a modified Enraf Nonius nitrogen gas-flow system. The cell parameters were determined from refinement by using centered angular positions of 25 reflections with $11 \le \theta \le 25°$. The profiles of the different reflections were measured using θ -2θ step scan method. A total of 3148 intensities were recorded up to two theta of 116°. Only selected reflections with significant intensity were collected for reflections in the [0°-58.85°] theta range. A number of 1736 independent reflections of which 1412 were used in the refinement procedures.

The H atoms were located by difference Fourier synthesis as described in section 2.1. The data reduction and error analysis were carried out by using the programs of Blessing (1989) [21]. The experimental details and crystal data are displayed in Table 6. The crystal structure has been deposited at the Cambridge Crystallographic Data

Centre with the deposition number CCDC 275138. These data can be obtained free of charge from The Cambridge Crystallographic Data Centre via *www.ccdc.cam.ac.uk/data_request/cif*.

Table 6. Experimental details

Crystal data	
Chemical formula	$C_6H_5NO_3$
Chemical formula weight	139.11
Cell setting	Monoclinic
Space group	$P2_1/n$
a (Å)	11.026(4)
b (Å)	6.736(1)
c (Å)	8.119(21)
β (°)	97.73 (2)
V (Å^3)	597.50
Z	4
D_x(mg m^{-3})	1.54
Radiation type	Mo $K\alpha$
Wavelength (Å)	0.71073
No. of reflections for cell parameters	25
θ range (°)	11 – 25
μ (mm^{-1})	0.127
Temperature (K)	122 (1)
Crystal form	Prism
Crystal size	$0.10 \times 0.27 \times 0.32$
Crystal color	Colorless
Data collection	
Diffractometer	Nonius CAD-4
Data collection method	$\theta \Box 2\theta$
$2\theta_{max}$	117.71
No. of measured reflections	3148
No. of independent reflections	1736
No. of observed reflections	1412
Criterion for observed reflections	$I \geq 3\sigma(I)$
R_{int}	0.021
Range of h, k, l	$-25 \rightarrow h \rightarrow 23$
	$0 \rightarrow k \rightarrow 15$
	$0 \rightarrow l \rightarrow 18$
No. of standard reflections	3
Frequency of standard reflections	Every 120 min
Refinement	

	N	R	wR	S
Spherical refinement	92	0.038	0.040	1.02
Multipole refinement	221	0.022	0.035	1.12

N is the number of refined parameters and M is the number of observations. $R = \sum |F_o| - |F_c| / \sum |F_o|$; $wR = \left[\sum w(|F_o| - |F_c|)^2 / \sum w|F_o|^2 \right]^{\frac{1}{2}}$; $S = \left[\sum w(|F_o| - |F_c|)^2 /(M-N) \right]^{\frac{1}{2}}$

Source of atomic scattering factors	*International Tables for X-ray Crystallography* (1999, Vol. C)

5. Conclusion

This study has obtained good accurate results on the structure and electron charge density which gives high quality descriptive model for the electron charge density distribution from X-ray diffraction experiment. It also revealed that electron density can yield to electronic proprieties such as dipole moment.

According to the orientation of the molecular dipole moment, the region of the nitro and hydroxyl groups is electronegative, whereas the region of the C-H groups is electropositive.

The coplanarity of the hydrogen bonded molecules together with the donor-acceptor interactions across the molecules probably enhance the nonlinear response of the orthorhombic mNPH, as it was discussed recently in case of the other nitrobenzene derivative [23]. In the monoclinic mNPH, the centrosymmetry of the crystal cancels the nonlinear response.

Finally, our results could be analyzed in more detail, if they were completed by quantum chemistry calculations. Especially, for the explanations about the existence of the polymorphism in m–Nitrophenol compounds.

References

1. Hamzaoui, F.; Baert, F.; Wojcik, G. Electron-density study of *m*-nitrophenol in the orthorhombic structure. *Acta Cryst. B* 1996, *52*, 159-164.
2. Wojcik, G.; Marqueton, Y. The phase transition of m-nitrophenol. *Mol. Cryst. Liq. Cryst.* 1989, *168*, 247-254.
3. Wojcik, G.; Toupet, L. The inter-and intramolecular charge transfer along the polymeric chain of hydrogen-bonded molecules in two crystal forms of m-nitrophenol. *Mol. Cryst. Liq. Cryst.* 1993, *229*, 153-159.
4. Panadares, F.; Ungaretti, L.; Coda, A. The crystal structure of a monoclinic phase of *m*-nitrophenol. *Acta Cryst. B* 1975, *31*, 2671-2675.
5. Koritsanszky, T.; Howard, S.; Richter, T.; Su, Z.; Mallinson, P.R.; Hansen, N.K. XD a Computer Program Package for Multipole Refinement and Analysis of Electron Densities from Diffraction Data. Free University of Berlin, Berlin, Germany, 2003.
6. Hansen, N.K.; Coppens, P. Testing aspherical atom refinements on small-molecule data sets. *Acta Cryst. A* 1978, *34*, 909-921.
7. Volkov, A.; Abramov, Y.; Coppens, P.; Gatti, C. On the origin of topological differences between experimental and theoretical crystal charge densities. *Acta Cryst. A* 2000, *56*, 332-339.
8. Volkov, A.; Gatti, C.; Abramov, Y.; Coppens, P. Evaluation of net atomic charges and atomic and molecular electrostatic moments through topological analysis of the experimental charge density. *Acta Cryst. A* 2000, *56*, 252-258.
9. Jeffrey, G.A.; Cruickshank, D.W.J. Molecular structure determination by *X*-ray crystal analysis: modern methods and their accuracy. *Quart. Rev. Chem.Soc.* 1953, *7*, 335-376.
10. *International Tables for X-ray Crystallography*, Kynoch Press, Birmingham, 1962, Vol.III.
11. Stewart, R.F.; Davidson, E.R.; Simpson, W.T. coherent X-Ray Scattering for the Hydrogen Atom in the Hydrogen Molecule. *J. Chem, Phys.* 1965, *42*, 3175-3187.
12. Coppens, P. X-ray Charge Densities and Chemical Bonding. New York: Oxford, 1997.
13. *International Tables for X-ray Crystallography*. Kluwer Academic Publishers, 1999, *Vol C*.

14. Trueblood, K.N. *Program THMAI1*, Department of chemistry and biochemistry, University of California, Los Angeles, 1990.
15. Hirshfeld, F.L. Can X-ray data distinguish bonding effects from vibrational smearing?. *Acta Cryst. A* 1976, *32*, 239-244.
16. Hirshfeld, F.L. Difference densities by least-squares refinement: fumaramic acid. *Acta Cryst B* 1971, *27*, 769.
17. Hirshfeld, F.L. Bonded-atom fragments for describing molecular charge densities. Theor.Chim Acta 1977, *44*, 129.
18. Hirshfeld, F.L.; Hope, H. An x-ray determination of the charge deformation density in 2-cyanoguanidine. *Acta Cryst. B* 1980, *36*, 406.
19. Johnoson, C.K. ORTEP program. Report ORNL-3794, Oak Ridge National Laboratory, Tennessee, 1965.
20. Blessing, R.H. DREAD - data reduction and error analysis for single-crystal diffractometer data. *J. Appl. Cryst.* 1989, *22*, 396-397.
21. Souhassou, M.; Blessing,R.H. Topological analysis of experimental electron densities. *J. Appl. Cryst* 1999, *32*, 210-217.
22. Souhassou, M.; Lecomte, C.; Blessing, R.H; Aubry, A.; Rohmer, M.M.; Wiest, R.; Benard, M.; Marraud, M. Electron distributions in peptides and related molecules.1. An experimental and theoretical study of N-acetyl-L-tryptophan methylamide. *Acta Cryst. B* 1991, *47*, 253-266.
23. Wojcik, G.; Holband, J.; Szymczak, J.J.; Roszak, S.; Leszczynski, J. Interactions in Polymorphic Crystals of m-Nitrophenol as Studied by Variable-Temperature X-ray Diffraction and Quantum Chemical Calculations. *Crystal Growth & Design* 2006, *6*, 274-282.

STRUCTURAL AND ELECTRON CHARGE DENSITY STUDIES IN A NONLINEAR OPTICAL COMPOUND 4,4 DI-METHYL AMINO CYANO BIPHENYL

Naima Boubegra, Abdelkader Chouaih*, Mokhtaria Drissi and Fodil Hamzaoui

Laboratoire de Structure, Elaboration et Applications des Matériaux Moléculaires (SEA2M), Université Abdelhamid Ibn Badis de Mostaganem, 27000 Mostaganem, Algeria.

Abstract:

The 4, 4 dimethyl amino -cyanobiphenyl crystal (DMACB) is characterized by its nonlinear activity. The intra molecular charge transfer of this molecule results mainly from the electronic transmission of the electro-acceptor (Cyano) and electro-donor (Di-Methyl-Amino) groups.

An accurate electron density distribution around the molecule has been calculated from a high-resolution X-ray diffraction study. The data were collected at 123 K using graphite-monochromated Mo-Kα radiation to $sin(\theta)/\lambda$ = 1.24 Å^{-1}. The integrated intensities of 13796 reflections were measured and reduced to 6501 independent reflections with $I \geq 3\sigma(I)$. The crystal structure was refined using the experimental model of Hansen and Coppens (1978). The crystal structure has been validated and deposited at the Cambridge Crystallographic Data Centre with the deposition number CCDC 876507.

A long this article, we present the thermal motion and the structural analysis obtained from the least square refinement based on F^2 and the electron density distribution obtained from the multipolar model.

Keywords: Electron charge density, DMACB, XD program, multipolar refinement

Introduction

In the nonlinear optical (NLO) field, the organic molecular crystals have attracted much interest, mainly because of their optical transparency and optical efficiency, which can be superior by several orders of magnitude to those of the best known mineral materials. Among them, organic second-order NLO molecules have been widely investigated by experimental and theoretical chemists as their high electro-optic (EO) coefficients, facile processing as well as low dielectric constants. [1–5] These molecules are prepared with suitable donor-acceptor (D, A) couples connected by a transmitter group T (π-electron). These kind of molecules are known under the term push-pull molecules.[6–10] A typical organic compound consists of strong electron acceptors (e.g. NO$_2$ or CN groups) and donors (e.g. NR$_2$ or OR groups) connected by a π-conjugated system. This arrangement ensures efficient intramolecular charge-transfer and enables further fine-tuning of the polarisability of the compound. [11–14]

The 4, 4 Di-Methyl Amino Cyano-Biphenyl compound (DMACB) is considered as one of the most interesting materials in non-linear optics because it crystallizes readily compared to other competitive crystals. The structural analysis has been already investigated at room temperature by Zyss and al. [15] DMACB crystallizes in the non-centrosymmetric space group Cc with four molecules in the unit cell. Its structure is shown in scheme 1.

The present work focuses on the thermal motion analysis and the electronic charge density distribution of the DMACB molecule using high-resolution X-ray diffraction data performed at 123 K. To achieve this study, we have used the XD program package based on the Hansen-Coppens charge density model.[16] One major component of this package is the program for the least squares fitting of the experimental data.[17]

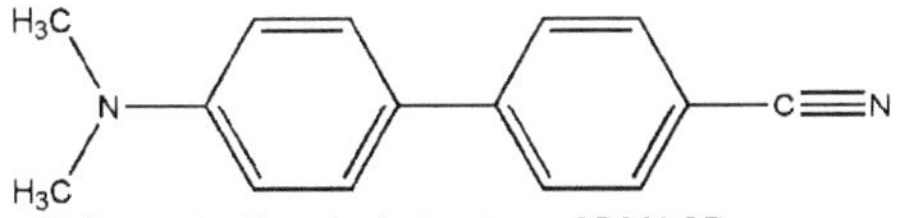

Scheme 1: Chemical structure of DMACB

Experimental

The crystallographic data were obtained at 123 K on a CAD4 diffractometer using the graphite–monochromated MoKα radiation. The crystal of high quality was cooled by using a modified Enraf Nonius nitrogen gas-flow system. A crystal of dimensions 0.2 × 0.2 × 0.15 *mm* was used for the data collection. The cell parameters were determined from refinement by using centered angular positions of 25 reflections with $11 \leq \theta \leq 25°$. The profiles of the different reflections were measured using θ–2θ step scan method. A total of 13796 intensities were recorded up to two theta of 59.7°. Only selected reflections with significant intensity were collected for the high order theta range. The data merged to give 6501 independent reflections. Only 5729 reflections for which $I \geq 3\sigma(I)$ were used for the refinement.

The H atoms were located by difference Fourier synthesis. The data reduction and error analysis were carried out by using the program of Blessing (1989).[18] The experimental details and crystal data are displayed in Table 1.

The crystal structure has been validated and deposited at the Cambridge Crystallographic Data Centre with the deposition number CCDC 876507. The full details can be obtained free of charge from The Cambridge Crystallographic Data Centre at www.ccdc.cam.ac.uk/data_request/cif.

Table 1. Experimental details

Crystal data	
Chemical formula	$C_{15}H_{14}N_2$
Chemical formula weight	222
Cell setting	Monoclinic
Space group	Cc
a (Å)	9.503 ± 2
b (Å)	16.429 ± 6
c (Å)	8.954 ± 4
β (°)	122.040 ± 3
V (Å^3)	1185.01 ± 8
Z	4
D_x(mg.m^{-3})	1.40 ± 01
Radiation type	MoKα
Wavelength (Å)	0.71073
No. of reflections for cell parameters	25
θ range (°)	2–29.6
μ (cm^{-1})	0,69
Temperature (K)	123 (1)
Crystal form	Prism
Crystal size (mm)	$0.2 \times 0.2 \times 0.15$
Crystal color	Coulorless
Data collection	
Diffractometer	Nonius CAD-4
Data collection method	$\theta - 2\theta$
$2\theta_{max}$	59.7
No. of measured reflections	13796
No. of independent reflections	6501
No. of observed reflections	5729
Criterion for observed reflections	$I \geq 3\sigma(I)$
R_{int}	0.018
Range of h, k, l	$0 \rightarrow h \rightarrow 11$
	$0 \rightarrow k \rightarrow 12$
	$-12 \rightarrow l \rightarrow 12$
No. of standard reflections	3
Frequency of standard reflections	Every 60 min

Table 2. Least square refinement factors

Refinement				
	N	R	wR	S
Spherical refinement	195	0.028	0.026	1.02

N is the number of refined parameters.

$$R = \sum |F_o| - |F_c| / \sum |F_o|; \quad wR = \left[\sum w(|F_o| - |F_c|)^2 / \sum w|F_o|^2 \right]^{1/2}; \quad S = \left[\sum w(|F_o| - |F_c|)^2 / (M - N) \right]^{1/2}$$

Source of atomic scattering factors	International Tables for X-ray Crystallography (1999, Vol. C) [19]

Refinement

In addition to conventional least-square refinement of data sets, a separate refinement using only high-order reflections ($sin\theta/\lambda \geq 0.75$ Å^{-1}) was performed an accurate processing using also the MOPRO program based on the Hansen-Coppens charge density model.

Structural analysis

In order to investigate the theoretical calculations of the molecular structure parameters, we have applied the ab intio methods. These calculations were performed with the Density Functional Theory (DFT) at B3LYP (Becke's three parameter hybrid functional using the correlation functional of Lee, Yang, and Parr, which includes both local and non-local terms correlation functional) method at 6-31G* level using the Gaussian 03 computational chemistry program package. [22]

The parameters of the optimized structure (bond lengths and bond angles) are listed in Tables 3 and 4. It can be seen that all the calculated parameters are in accordance with the X-ray results. In spite of the differences, calculated geometric parameters represent a good approximation and can provide a starting point to calculate other parameters, such as vibrational wavenumbers, as will be described in the next work.

It is worth noting that some of the optimized bond lengths and bond angles have slightly different values from the corresponding experimental ones, due to the fact that the theoretical calculations consider only isolated molecules in the gaseous phase while the experimental results refer to molecules in the crystal environment.

The general features of the structure have been described previously at room temperature. The DMACB is a non-planar molecule. The value of the dihedral angle between the two phenyl groups is 42°. The main characteristic of this structure is that the four molecules in the unit cell are engaged in four chains formed by infinite chains of hydrogen-bonded coplanar molecules as shown in Figure 1. Our results have not been subject to any structural changes compared to the initial published results.

The design of most efficient organic materials for the non-linear optical (NLO) effect is based on molecular units containing highly delocalized pi-electron moieties and extra electron donor (D) and electron acceptor (A) groups on opposite sides of the molecule at appropriate positions on the ring to enhance the conjugation. The pi-electron cloud movement from donor to acceptor makes the molecule to be highly polarized. DMACB molecule was designed by using acceptor group N≡C– (cyano group) which has the highest acceptor ability, and the donor group –N(CH₃)₂ (dimethyl amino group) which has the highest donor character. The ab initio optimization in this investigation shows that the molecule of DMACB is almost planer. The planarity can affect the NLO properties of DMACB due to the free rotation of two benzene rings. The DMACB molecule has two benzene rings connected through a single covalent bond C–C, which can eventually stop the free rotation and facilitate the intramolecular charge transfer. To understand this phenomenon in the context of molecular orbital picture, molecular HOMOs and molecular LUMOs can be examined.

Thermal motion analysis

The thermal motion analysis of DMACB has been performed using the THMA11 program. [23] The rigid-body motion is described by three tensors T for the translation, L for the liberation and S for taking into account the correlation between translation and liberation. These tensors are obtained by a least-square fit refinement using the observed atomic thermal motion parameters obtained by the refinement and given in Table 5. This test indicates that the differences between the mean square displacement amplitudes (MSDAs) along the interatomic directions have magnitude $\Delta \leq 10.10^{-4} \text{Å}^2$ for the most bonded pairs of the non-H atoms. [24] The MSDAs Δ_{AB} in the AB direction for all pairs of atoms in the molecule have been calculated and reported in Table 6.

In the general treatment of the molecular thermal motion in terms of rigid-body (TLS), the calculated anisotropic thermal parameters are given in the Trueblood notation such as:

$$U_{ij} = T_{ij} + G_{ijkl}L_{kl} + H_{ijkl}S_{kl} + D^2\Omega^2 n_i n_j$$

(2)

where G, H and D are geometrical parameters. The last term corresponds to any additional intra-libration (Ω) around a chosen axis. The rigid-body fit suggests two independents liberation axis around the C(10)–N(2) and C(7)–C(6) bonds. The thermal motion of the H atoms is considered to consist of two contributions. The first is due to rigid molecular motion and the second is from the C–H vibrations. [25-27] Values of the T, L and S tensors obtained from the least square fitting are given in Table 7.

The C–H bond frequencies of the molecule were taken from Baert et al. [28] The corresponding mean-square displacements for the H atoms in phenyl group were 0.0056, 0.014 and 0.025 A² for bond stretching, in-plane bending and out-of-plane bending respectively. While, for the H atoms of the methyl group the internal vibration amplitudes used were 0.0057, 0.0116 and 0.0224 A². The results of the thermal motion parameters of the H atoms are summarized in Table 8. The ellipsoids of the different atoms representing their thermal motion described above are shown, using ORTEPIII [29] diagram, in Figure 2.

Table 3. Bond lengths (Å) of DMACB, with estimated deviations in parentheses

Atom 1	Atom 2	Distance		Atom 1	Atom 2	Distance	
		X-ray	DFT			X-ray	DFT
N(1)	C(13)	1.144 (3)	1.137	C(4)	C(5)	1.374 (3)	1.381
N(2)	C(10)	1.372 (3)	1.410	C(5)	C(6)	1.401 (3)	1.394
N(2)	C(14)	1.435 (3)	1.453	C(6)	C(7)	1.468 (3)	1.488
N(2)	C(15)	1.436 (4)	1.445	C(7)	C(8)	1.393 (4)	1.395
C(1)	C(2)	1.377 (3)	1.381	C(7)	C(12)	1.393(3)	1.386
C(1)	C(6)	1.397 (4)	1.394	C(8)	C(9)	1.382 (3)	1.377
C(2)	C(3)	1.384 (4)	1.390	C(9)	C(10)	1.393 (3)	1.398
C(3)	C(4)	1.391 (4)	1.390	C(10)	C(11)	1.394 (4)	1.387
C(3)	C(13)	1.438 (3)	1.444	C(11)	C(12)	1.382 (3)	1.391

Table 4. Bond angles (deg) of DMACB, with estimated deviations in parentheses

Atom 1	Atom 2	Atom 3	Angle		Atom 1	Atom 2	Atom 3	Angle	
			X-ray	DFT				X-ray	DFT
C(10)	N(2)	C(14)	120.9 (3)	116.0	C(1)	C(6)	C(5)	117.6 (2)	116.0
C(10)	N(2)	C(15)	121.2 (2)	117.3	C(1)	C(6)	C(7)	121.1 (2)	119.3
C(14)	N(2)	C(15)	117.9 (2)	112.3	C(5)	C(6)	C(7)	121.2 (2)	121.3
C(2)	C(1)	C(6)	121.4 (2)	121.0	C(6)	C(7)	C(8)	121.6 (2)	121.0
C(1)	C(2)	C(3)	119.9 (2)	120.1	C(6)	C(7)	C(12)	121.8 (3)	119.9
C(2)	C(3)	C(4)	119.9 (2)	121.0	C(8)	C(7)	C(12)	116.6 (2)	116.8
C(2)	C(3)	C(13)	120.2 (3)	118.4	C(7)	C(8)	C(9)	121.9 (2)	120.1
C(4)	C(3)	C(13)	119.8 (2)	120.8	C(8)	C(9)	C(10)	121.3 (3)	121.0
C(3)	C(4)	C(5)	119.8 (2)	120.8	N(2)	C(10)	C(9)	121.5 (3)	120.8
C(4)	C(5)	C(6)	121.3 (2)	117.4	N(2)	C(10)	C(11)	121.6 (2)	121.2
C(7)	C(12)	C(11)	121.8 (3)	119.2	C(9)	C(10)	C(11)	116.9 (2)	117.0
N(1)	C(13)	C(3)	180.0 (1)	180.0	C(10)	C(11)	C(12)	121.4 (2)	120.4

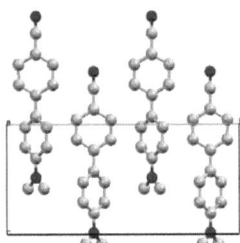

Fig. 1. Packing diagram of the structure.

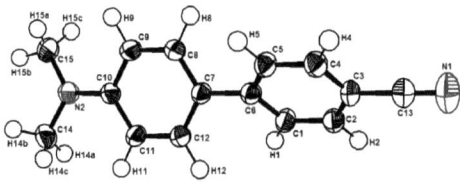

Fig. 2. ORTEPIII Diagram of the DMACB molecule with atomic labeling scheme.

Table 5. Atomic displacement parameters (Å^2)

Atoms	U_{11}	U_{22}	U_{33}	U_{12}	U_{13}	U_{23}
N(1)	0.0331 (10)	0.0392 (11)	0.0148 (7)	0.0026 (8)	0.0081 (7)	−0.0006 (7)
N(2)	0.0164 (7)	0.0286 (9)	0.0102 (5)	−0.0055 (5)	0.0053 (5)	−0.0005 (5)
C(1)	0.0176 (7)	0.0156 (7)	0.0127 (6)	−0.0006 (5)	0.0077 (5)	0.0008 (5)
C(2)	0.0169 (7)	0.0155 (7)	0.0143 (6)	−0.0014 (5)	0.0070 (6)	0.0003 (5)
C(3)	0.0164 (7)	0.0187 (7)	0.0103 (6)	−0.0002 (5)	0.0056 (5)	−0.0002 (5)
C(4)	0.0174 (7)	0.0164 (7)	0.0134 (6)	−0.0020 (5)	0.0071 (5)	−0.0019 (5)
C(5)	0.0183 (7)	0.0143 (6)	0.0131 (6)	−0.0021 (5)	0.0075 (5)	−0.0005 (5)
C(6)	0.0148 (7)	0.0144 (7)	0.0128 (6)	−0.0010 (5)	0.0073 (5)	−0.0002 (5)
C(7)	0.0149 (7)	0.0137 (6)	0.0120 (6)	−0.0012 (5)	0.0068 (5)	0.0000 (5)
C(8)	0.0142 (6)	0.0202 (7)	0.0134 (6)	−0.0034 (5)	0.0067 (5)	0.0001 (5)
C(9)	0.0161 (7)	0.0182 (7)	0.0139 (6)	−0.0032 (5)	0.0083 (5)	−0.0001 (5)
C(10)	0.0134 (6)	0.0151 (6)	0.0123 (5)	−0.0006 (4)	0.0062 (5)	−0.0005 (5)
C(11)	0.0170 (7)	0.0193 (7)	0.0120 (6)	−0.0034 (5)	0.0075 (5)	0.0006 (5)
C(12)	0.0155 (7)	0.0186 (7)	0.0131 (6)	−0.0035 (5)	0.0068 (5)	0.0001 (5)
C(13)	0.0192 (8)	0.0200 (8)	0.0144 (7)	0.0024 (6)	0.0075 (6)	0.0009 (6)
C(14)	0.0159 (7)	0.0278 (8)	0.0139 (6)	−0.0018 (6)	0.0056 (5)	−0.0005 (6)
C(15)	0.0203 (7)	0.0271 (8)	0.0140 (6)	−0.0010 (6)	0.0101 (6)	0.0017 (6)

Table 6. Matrix for differences in MSDA's (mean square displacements of atoms). Values listed are 10^4 MSDA's for column atom minus that for row atom, underlined values correspond to chemical bonds.

	C15	C14	C13	C12	C11	C10	C9	C8	C7	C6	C5	C4	C3	C2	C1	N2
N1	−4	4	**13**	−4	−6	−12	−4	−6	2	−7	−16	−17	−12	−15	−4	−16
N2	1	**22**	29	4	5	**4**	28	10	19	9	26	8	5	9	16	
C1	−5	−1	15	−4	13	−17	−7	11	−8	**4**	−9	2	−6	**7**		
C2	2	9	20	5	−2	−9	1	13	3	−3	−17	15	**8**			
C3	13	17	**25**	12	9	−1	10	12	14	4	−10	**8**				
C4	12	14	−2	15	10	−4	7	8	12	15	**10**					
C5	−4	−4	15	3	−11	−24	−10	−7	−14	**9**						
C6	8	10	20	4	0	−5	5	19	**10**							
C7	−5	1	11	**5**	−6	−14	−10	**17**								
C8	10	12	14	−12	−2	−5	**6**									
C9	5	−7	15	5	3	**13**										
C10	−6	16	25	−4	**4**											
C11	−8	11	16	**1**												
C12	1	9	13													
C13	−12	−6														
C14	−7															

Table 7. Rigid body vibration parameters of the T, L and S tensors

T (rad^2)

$$\begin{pmatrix} 0.00136 & & \\ -0.00006 & 0.00037 & \\ -0.00008 & -0.00005 & 0.00044 \end{pmatrix}$$

L (Å^2)

$$\begin{pmatrix} 0.01089 & & \\ -0.00203 & 0.01029 & \\ -0.00031 & -0.00013 & 0.01289 \end{pmatrix}$$

S (rad Å)

$$\begin{pmatrix} 0.00029 & -0.00001 & 0.00000 \\ 0.00015 & -0.00038 & 0.00010 \\ 0.00014 & 0.00003 & 0.00000 \end{pmatrix}$$

Table 8. The thermal vibration parameters for the H atoms calculated from the TLS group tensors plus internal C–H bond vibration

Atoms	U_{11}	U_{22}	U_{33}	U_{12}	U_{13}	U_{23}
H(1)	0.0307	0.0134	0.0163	0.0005	0.0058	0.0009
H(2)	0.0352	0.0188	0.0163	0.0002	0.0072	0.0010
H(4)	0.0322	0.0190	0.0138	-0.0035	0.0047	-0.0050
H(5)	0.0255	0.0141	0.0136	-0.0026	0.0035	-0.0038
H(8)	0.0150	0.0272	0.0130	-0.0061	0.0032	-0.0016
H(9)	0.0212	0.0307	0.0131	-0.0059	0.0040	-0.0011
H(11)	0.0188	0.0273	0.0183	-0.0085	0.0012	-0.0049
H(12)	0.0180	0.0261	0.0183	-0.0091	0.0045	-0.0035
H(14A)	0.0154	0.0405	0.0189	-0.0004	-0.0029	-0.0049
H(14B)	0.0310	0.0239	0.0173	-0.0072	-0.0010	-0.0064
H(14C)	0.0219	0.0265	0.0160	0.0016	-0.0029	-0.0049
H(15A)	0.0257	0.0393	0.0132	0.0025	0.0058	-0.0018
H(15B)	0.0321	0.0271	0.0130	-0.0038	0.0032	-0.0008
H(15C)	0.0292	0.0263	0.0128	0.0034	0.0028	-0.0027

Electron-density maps:

The aspherical atom model used in multipolar refinement gives structure factor phases closer to the true phases for crystals than the spherical or independent atom model does. [30-31] This enables the mapping of the electron density by Fourier synthesis in various ways using the program XDGRAPH implemented in the XD program package. [15]

The experimental density deformation map is shown in Figure 3, from which we can notice the absence of the density on the atomic sites and the appearance of all the bond density peaks. This map confirms the high quality of the data sets and the efficiency of the formalism used for the data processing as proposed by Blessing. [18] This visualization is obtained using the calculated multipole phases with the observed structure factors F$_{obs}$ (h):

$$\delta\rho^{exp}(r) = \frac{1}{V}\sum_h \left[|F_{Obs}(h)|.e^{i\phi_{mul}} - |F_{sph}(h)|.e^{i\phi_{sph}} \right] e^{-2\pi i h.r} \tag{3}$$

$F_{sph}(h)$ is computed with atomic positions and thermal parameters. Experimental density map from high-order refinement: $\rho_{exp} = \rho_o - \rho_{sph}$. Where, ρ_o is the observed electron density and ρ_{sph} is the calculated electron density using the atomic parameters obtained from the high-order refinement.

We have explored four planes to visualize the electron density distribution: the plane of the aromatic cycle containing the C(1), C(2), C(3), C(4), C(5) and C(6) atoms; the plane of the second aromatic cycle containing the C(7), C(8), C(9), C(10), C(11) and C(12) atoms; the third plane formed by the elctro-donor group (amino dimethyl); the last plane containing the elctro-acceptor goup (cyano). Figure 3 gives the different maps cited above, all contours map are 0.05 e.Å^{-3}.

We observe that the electron density distribution is almost centered in the middle of the chemical bonds. We also notice that the peaks (Figure 3d) of electron density in the connection N(2)−C(14) and N(2)−C(15) are centered towards the nitrogen N(2) backing the electro donor character of the methyl group. The symmetrical electron distribution along the both sides of C(13)−N(1) axis shows clearly the multi bonding aspect of the C(13)−N(1) chemical bond.

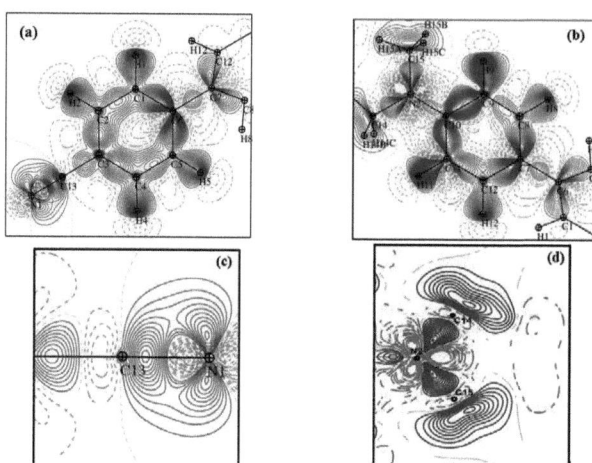

Fig. 3. Deformation density map with contour map of 0.05 eÅ^{-3}. Positive density in red and negative density in blue : (a) the plane of first aromatic cycle ; (b) the plane of second aromatic cycle ; (c) the plane of the cyano group ; (d) the plane of dimethyl amino group.

Conclusion

The ab initio theoretical structure investigation leads to an optimized molecule with bond lengths and bond angles closer to the X-ray experiment results. No significant structural changes were observed compared to the previous results obtained at room temperature. However, the thermal motion analysis using the Hirshfeld rigid-bond test shows the advantages of the very low temperature data collection. The atomic motions have been significantly reduced. The maximum discrepancy is only 0.0005 A^2 after the multipolar refinement.

The multipolar model of Hansen and Coppens allowed us to obtain the electron charge density function. In order to check the quality of our analysis, we have used the obtained density function to visualize the electron charge density distribution around the different atoms.

We have explored systematically the main molecule planes. The different sections show clearly the accumulation of the electron charge density along the chemical bonding. The oxygen lone pairs have been perfectly localized.

The good quality of the obtained results can only confirm the accuracy of the obtained charge density function which will lead us in our next step in a forthcoming paper to the calculation of the molecular dipole moment and the electrostatic potential around the molecule in order to confirm the nature of electron charge transfer of the DMACB compound.

References

[1] Bredas J L, Adant C, Tackx P, Persoons A and Pierce B M 1994 Chem. Rev. **94** 243
[2] He M, Leslie T M, Sinicropi J A, Garner S M and Reed L D 2002 Chem. Mater. **14** 4669
[3] Locatelli D, Quici S, Roberto D and De Angelis F 2005 Chem. Commun. **43** 5405
[4] Kang H, Facchetti A, Jiang H, Cariati E, Righetto S, Ugo R, Zuccaccia C, Macchioni A, Stern C L, Liu Z, Ho S T, Brown E C, Ratner M A and Marks T J 2007 J. Am. Chem. Soc. **129** 3267
[5] Hu Y Y, Sun S L, Zhong R L, Xu H L and Su Z M 2011 J. Phys. Chem. C **115** 18545
[6] Zyss J 1994 Molecular Nonlinear Optics: Materials, Physics and Devices (New York: Academic Press)
[7] Nalwa H, Miyata S 1996 Nonlinear Optics of Organic Molecules and Polymers (New York: CRC Press)
[8] Ledoux I, Zyss J 1997 Molecular nonlinear optics: fundamentals and applications, Chapter I, in: Khoo I C, Simoni F, Umeton C (Eds.), Novel Optical Materials and Applications, pp. 1–48.
[9] Andraud C, Zabulon T, Collet A and Zyss J 1999 Chemical Physics **245** 243
[10] Zhou Z J, Li X P, Ma F, Liu Z B, Li Z R, Huang X R and Sun C C 2011 Chem. Eur. J. **17** 2414
[11] Bureš F, Schweizer W B, May J C, Boudon C, Gisselbrecht J P, Gross M, Biaggio I and Diederich F 2007 Chem Eur J **13(19)** 5378
[12] Bureš F, Schweizer W B, Boudon C, Gisselbrecht J P, Gross M and Diederich F 2008 Eur J Org Chem **6** 994
[13] May J C, Biaggio I, Bureš F and Diederich F 2007 Appl Phys Lett. **90** No. 251106
[14] Bureš F, Pytela O and Diederich F 2009 J Phys Org Chem **22(2)** 155
[15] Zyss J, Ledoux I, Bertault M and Toupet E 1991 Chemical Physics **150** 125
[16] Hansen N K and Coppens P 1978 Acta Cryst. A **34** 909
[17] Koritsanszky T, Howard S, Richter T, Su Z, Mallinson P R and Hansen N K 2003 XD a Computer Program Package for Multipole Refinement and Analysis of Electron Densities from Diffraction Data (Free University of Berlin, Berlin, Germany)
[18] Blessing R H 1989 J. Appl. Cryst. **22** 396
[19] International Tables for X-ray Crystallography Vol C 1999 (Kluwer Academic Publishers)
[20] Stewart R F, Davidson E R and Simpson W T 1965 J. Chem, Phys. **42** 3175
[21] Becker P J and Coppens P 1974 Acta Cryst. A **30** 129
[22] Frisch M J, Trucks G W, Schlegel H B, Scuseria G E, Robb M A, Cheeseman J R, Montgomery J A, Jr, Vreven T, Kudin K N, Burant J C, Millam J M, Iyengar S S, Tomasi J, Barone V, Mennucci B, Cossi M, Scalmani G, Rega N, Petersson G A, Nakatsuji H, Hada M, Ehara M, Toyota K, Fukuda R, Hasegawa J, Ishida M, Nakajima M, Honda Y, Kitao O, Nakai H, Klene M, Li X, Knox J E, Hratchian H P, Cross J B, Adamo C, Jaramillo J, Gomperts R, Stratmann R E, Yazyev O, Austin A J, Cammi R, Pomelli C, Ochterski J W, Ayala P Y, Morokuma K, Voth G A, Salvador P, Dannenberg J J, Zakrzewski V G, Dapprich S, Daniels A D, Strain M C, Farkas O, Malick D K, Rabuck A D, Raghavachari K, Foresman J B, Ortiz J V, Cui Q, Baboul A G, Clifford S, Cioslowski J, Stefanov B B, Liu G, Liashenko A, Piskorz P, Komaromi I, Martin R L, Fox D J, Keith T, Al-Laham M A, Peng C Y, Nanayakkara A, Challacombe M, Gill P, M W, Johnson B, Chen W, Wong M W, Gonzalez C and Pople J A 2003 Gaussian 03 (Pittsburgh PA, Gaussian 03 Revision A.1)
[23] Trueblood K N 1990 Program THMAI1 (Department of chemistry and biochemistry, University of California, Los Angeles)
[24] Hirshfeld F L 1976 Acta Cryst. A **32** 239
[25] Hirshfeld F L 1971 Acta Cryst. B **27** 769
[26] Hirshfeld F L 1977 Theor. Chim Acta 44 129
[27] Hirshfeld F L, Hope H 1980 Acta Cryst. B **36** 406
[28] Baert F, Schweiss P, Heger G and More M 1988 J. Mol. Struct. **178** 29
[29] Johnoson C K 1965 ORTEP Program Report ORNL-3794 (Tennessee: Oak Ridge National Laboratory)
[30] Coppens P 1997 X-ray Charge Densities and Chemical Bonding (New York: Oxford)
[31] Souhassou M and Blessing R H 1999 J. Appl. Cryst. **32** 210

Theoretical and Experimental Electrostatic Potential Around the m-Nitrophenol Molecule

Mokhtaria Drissi, Nadia Benhalima, Youcef Megrouss, Rahmani Rachida, Abdelkader Chouaih and Fodil Hamzaoui

Abstract:

This work conerns a comparison of experimental and theoretical results of the electron charge density distribution and the electrostatic potential around the *m*-nitrophenolmolecule (m-NPH)known for its interesting physical characteristics. The molecular experimental results have been obtained from a high-resolution X-ray diffraction study. Theoretical investigations were performed using the Density Functional Theory at B3LYP level of theory at 6-31G* in the Gaussian program. The multipolar model of Hansen and Coppens was used for the experimental electron charge density distribution around the molecule, while we used the DFT methods for the theoretical calculations. The electron charge density obtained in both methods allowed us to find out different molecular properties such us the electrostatic potential and the dipole moment, which were finally subject to a comparison leading to a good match obtained between both methods. The intramolecular charge transfer has also been confirmed by an HOMO-LUMO analysis.

Keywords:

electron charge density; *m*-nitrophenol; nonlinear optical compound (NLO); electrostatic potential; optimized geometry; HOMO-LUMO

1.Introduction

m-Nitrophenol(m-NPH) occurs in two polymorphic forms: orthorhombic ($P2_12_12_1$) and monoclinic ($P2_1/n$) (see Figure 1). We are going to concentrate on the monoclinic form as the first form has already beenthe subject of a preceding report [1]. The main purpose of our work is to establish the electrostatic potential around the molecule through the determination of the electron charge density. This electrostatic potential will help us to describe and understand theinter- and intramolecular interactions (charge transfer) in the crystal. The presented electrostatic potential can be a starting point for the estimation of crystal energy cohesion in order to get more information about the existence of the polymorphism in the compound*m*-nitrophenol[2–4].

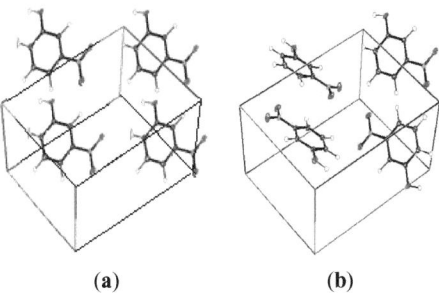

(a)   (b)

Figure 1. Polymorphic forms of *m*-nitrophenol, (**a**) orthorhombic, (**b**) monoclinic.

We have previously published an article about the high resolution X-ray diffraction and crystallographic study witha thermal motion analysis of the compound m-NPH[5]. The lasting metastability of the monoclinic form of *m*-nitrophenoloriginates probably from interactions within the centrosymmetric dimers of overlapping molecules. The attractive interactions within centrosymmetric dimers of overlapping molecules are as strong as interactions between hydrogen-bonded molecules [6,7].We also presented the molecular dipole moment based on spherical model refinement [8].In the present work, we based our study on the multipolar model of Coppens–Hansen where the non-spherical aspect of the atoms was taken into account. The multipole model represents an extrapolation to infinite resolution from a finite set of experimental data [9]. This last refinement gave us an accurate picture of the electronic charge density distribution in the compound m-NPH. The theoretical part adds to our experimental work by using *ab-initio* calculations through providing a comparison of the molecular electrostatic properties such as the dipole moment, the electron density maps and the electrostatic potential with the experimental data.

2. Crystallographic Details

The accurate electron density distribution and the electrostatic potential around the molecule (m-NPH) have been calculated from a high-resolution X-ray diffraction study [5]. A summarized table of the X-ray experiment details of the crystallographic data is given in Table 1.

Table 1. Experimental details.

Crystal Data	
Chemical formula	$C_6H_5NO_3$
Chemical formula weight	139.11
Cell setting	Monoclinic
Space group	$P2_1/n$
a (Å)	11.026(4)
b (Å)	6.736(1)
c (Å)	8.119(21)
$\beta(°)$	97.73 (2)
V (Å^3)	597.50
Z	4
Radiation type	Mo $K\alpha$
Temperature (K)	122 (1)
No. of measured reflections	3148

3. Computational Details

The theoretical calculations were performedusingthe Density Functional Theory at B3LYP (Becke's three parameter hybrid functional using the correlation functional of Lee, Yang, and Parr, which includes both local and non-local terms correlation functional) methods at 6-31G* level[15].

To perform this computational work, we used the Gaussian 09 program package [16].The Highest Occupied Molecular Orbital(HOMO–Lowest Unoccupied Molecular Orbital (LUMO) analysis has been carried out to explain the charge transfer place within the molecule. The chemical hardness and chemical potential are also calculated using the HOMO and LUMO.

The visualization of the electron charge and the electrostatic potential of the molecule were obtained using the Molden program where the Molden is a package for displaying molecular density from the *ab initio* package Gaussian [17].

4. Results and Discussion

4.1. Optimization of Geometrical Parameters

Geometry optimization is a name for the procedure that attempts to find the configuration of minimum energy of the molecule. The procedure calculates the wave function and the energy at a starting geometry and then proceeds to search a new geometry of a lower energy.

The optimized structure of the title compound isshown in Figure 2. The calculated structure parameters (bond lengths, bond angles and torsion angles) were listed inTables 2–4 where it can be seen that all the calculated parameters are in line with the X-ray results. In summary, the optimized bond lengths and bond angles obtained using the DFT method are in good agreement with the corresponding X-ray structural parameters. The calculated geometric parameters represent a good approximation and can provide a starting point to calculate other parameters, such as vibrational wavenumbers.

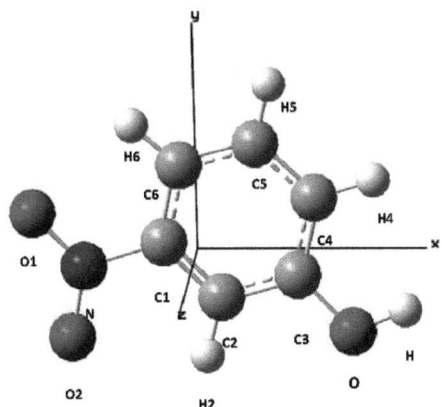

Figure 2. The optimized structure of m-NPH based on DFT B3LYP/6-1G* basis set.

Table 2. Selected bond distances (Å) by X-ray and theoretical calculations (B3LYP/6-31G*).

Atom 1	Atom 2	Distance(Å)	
		X-Ray	B3LYP/6-31G*
C1	C6	1.410	1.433
C1	C2	1.396	1.384
C2	C3	1.402	1.396
C3	C4	1.411	1.392
C4	C5	1.396	1.391
C6	C5	1.400	1.412
O	C3	1.365	1.380
C1	N	1.474	1.468
O1	N	1.244	1.281
O2	N	1.243	1.283
O	H	1.030	0.992
H6	C6	1.089	1.084
H2	C2	1.085	1.078
H4	C4	1.078	1.069
H5	C5	1.085	1.082

Table 3. Selected bond angles (°) by X-ray and theoretical calculations (B3LYP/6-31G*).

Atom 1	Atom 2	Atom 3	Angle (°)	
			X-Ray	B3LYP/6-31G*
C6	C1	N	118.90	118.46
C6	C1	C2	124.11	122.55
N	C1	C2	116.99	117.37
H2	C2	C3	122.50	119.45
H2	C2	C1	120.51	120.82
C3	C2	C1	118.99	119.10
C4	C3	O	123.31	122.72
H4	C4	C3	120.31	119.88
H4	C4	C5	118.98	118.94
C3	C4	C5	120.71	118.06
H5	C5	C6	119.20	119.87
H5	C5	C4	120.66	119.74
C6	C5	C4	120.14	119.28
H6	C6	C5	119.39	120.13
H6	C6	C1	123.00	120.91
C5	C6	C1	117.60	119.28
H	O	C3	109.00	110.55
O2	N	O1	123.70	122.07
O2	N	C1	119.09	117.04
O1	N	C1	117.20	117.03

Table 4. Torsion angles (°) by X-ray and theoretical calculations (B3LYP/6-31G*).

Atom1	Atom 2	Atom 3	Atom 4	Angle (°)	
				X-Ray	B3LYP/6-31G*
C4	C3	C2	C1	−1.07	0.003
C5	C6	C1	C2	0.92	−0.015
C4	C3	C2	C1	0.08	0.011
H2	C2	C3	C4	178.76	179.99
H6	C6	C1	C2	−179.28	−180.00
H5	C5	C6	C1	178.63	179.98
H4	C4	C3	C2	−179.68	−179.99
O	C3	C4	C5	−179.68	−179.97
H	O	C3	C4	−6.70	−179.98
N	C1	C2	C3	−179.84	0.011
O1	N	C1	C2	179.21	179.84
O2	N	C1	C2	0.39	0.020

4.2. Electron Density Maps

Figure 3provides a comparison of the experimental static charge density of the molecule, obtained by convolution of the thermal motion from the charge density on the different atoms in the mean molecular plane, with the theoretical charge density, determined from a wave function for a pseudo atoms from an ab initiocalculation performed with a Gaussian basis setusing the Density Functional Theory at B3LYP level of theory at 6-31G*. As it can be seen, the two maps show reasonable agreement. These maps confirm the high quality of the data sets and the efficiency of the formalism of data processing as proposed by Blessing [18].

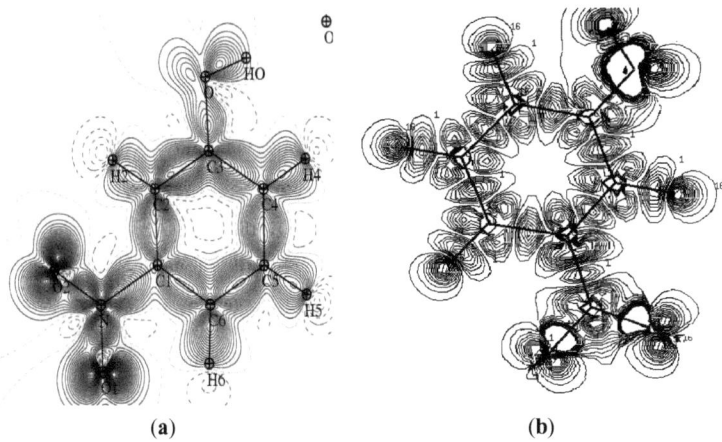

(a)   (b)

Figure 3. Comparison of the static and theoretical density maps of mNPH. (**a**) Static density map. (**b**) Theoretical electron density map.

4.3. Net Atomic Charges

Thevalence and mulitipolarpopulationceofficients were used to estimate the partial charges on the different atoms and the molecular dipole moment following the procedure described by Hansen and Coppens [8].The experimantal net atomic chargeshave been previously published in an article about the high resolution X-ray diffraction and crystallographic study of m-NPH [5]. These values arecompared to the natural population analysis (NPA) charges derived from the *ab initio*calculations using B3LYP with the 6-31G* basis set(see Table 5, Figure 4). All the methods are in agreement for the evaluation of the positive sign of the net charges on the H and N atoms and the negative net charges on the O atoms.

Table 5. Atomiccharge of *m*-nitrophenol.

Atom	Multipolar Refinement	B3LYP/6-31G*
C1	−0.1536	0.07099
C2	−0.2703	−0.26992
C3	0.0288	0.33201
C4	−0.3958	−0.29354
C5	−0.4621	−0.21591
C6	−0.2689	−0.24740
N	0.6466	0.51462
O1	−0.2337	−0.38131
O2	−0.2189	−0.38013
O	−0.2901	−0.68228
H2	0.2187	0.28515
H4	0.2165	0.24054
H5	0.2508	0.25207
H6	0.2885	0.27564
H	0.3289	0.49644

4.4. Molecular Moments

From the knowledge of the density function one can derive some important physical properties of the molecules such as the surrounding electrostatic field gradient, and the different electrostatic moments of the charge distribution [14]. A property associated to the average value of a quantum observable $\langle O \rangle$ is linked to the density function as given by the general equation (3), V is the molecular volume:

$$\langle O \rangle = \int_V \hat{O}(\vec{r})\rho(\vec{r})d\vec{r} \tag{1}$$

If $\Delta\rho(\vec{r})$ rather than $\rho(\vec{r})$ is being considered the electrostatic moment due to the deformation density in the molecule and can be estimated. The experimental molecular dipole moment of m-NPH has been determined inthe previous paper cited above using the multipolar model [5].Such studies have clearly evidenced the electron donor character of the C-H groups in conjunction with the electron acceptor character of the nitro and hydroxyl groups. In general, the experimental method provides a magnitude of about 5.80 Debyefor the dipole moment. A theoretical calculation has been performed usingB3LYP at 6-31G* basis set in order to carried out the components of the molecular dipole moment. The obtained results are summarized in Table 6 in which the experimental values are given for comparison.The orientation of the different vectors ofdipole moment in the molecular axial system is shown in Figure 5.

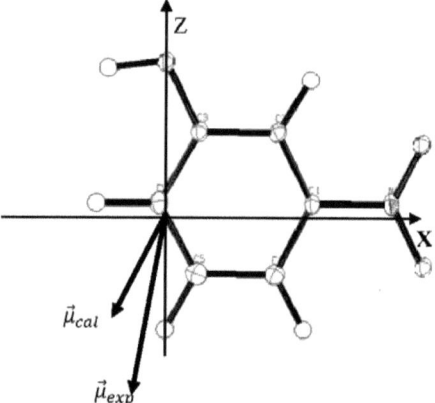

Figure 5. Orientation of the molecular dipole moment of m-NPH:$\vec{\mu}_{exp}$: molecular dipole moment from the experimental study;$\vec{\mu}_{cal}$: molecular dipole moment from the theoretical DFT calculations.

Table 6. Components of the molecular dipolar moment from DFT calculations (B3LYP at 6-31G* basis set) and x-ray experiment. The origin coincides with the center of mass of the molecule, and the Cartesian system referred to the inertial axis of the molecule.

Methods	Models	μ_X	μ_Y	μ_Z	μ Debye
X-rayExperiment	Multipolar refinement	−0.3209	−0.3200	−6.3358	5.8000
Ab initio	DFT(B3LYP/6-31G*)	−2.1194	−0.0010	−5.4234	5.8228

The components of the electrostatic quadrupole moment are obtained by substituting in Equation (3) the operator $\hat{O}(r)$ by $\vec{r}_\alpha \Lambda \vec{r}_\beta$. If in that equation the density function $\rho(\vec{r})$ is replaced by the multipolar expansion up to order $l = 1$, then the components of the quadrupole moment are given by:

$$Q_{\alpha\beta} = \sum \left[Q_{\alpha\beta}^i + r_{i\beta}d_{i\alpha} + r_{i\alpha}d_{i\beta} + r_{i\alpha}r_{i\beta}q_i \right] \tag{2}$$

where $d_{i\alpha}$ and q_i represent respectively the component of the dipole moment and the net charge of atom i at r_i. $Q_{\alpha\beta}^i$ are the atomic quadrupoles neglected here.

In the case of the direct integration method the development of Equation (3) leads to:

$$Q_{\alpha\beta} = \frac{1}{V} \sum_{\vec{H}} \Delta F(\vec{H}) \left[\sum_i \left(Q_{\alpha\beta}^i + r_{i\beta}d_{i\alpha} + r_{i\alpha}d_{i\beta} + r_{i\alpha}r_{i\beta}q_i \right) \right] \tag{3}$$

with:

$$Q_{\alpha\beta}^i = \int_{t_i} (r_\alpha - r_i)(r_\beta - r_i) e^{i2\pi\vec{H}(\vec{r}-\vec{r}_i)} d^3r \tag{6}$$

The summation over $\vec{H}$ is performed over all structure factors and the indice t_i designates the integrable subunits. Evaluation of all molecular moments requires summations of the density and moments of each subunit which are being performed according to a space partitioning scheme. The quadrupolar moment values are reported in the Table 7 with the analogous components obtained from the point charge model using the net atomic charges derived by NPA method calculations. The most remarkable features when comparing experimental values with those derived from the free molecule stand-out in the Q_{XX}, Q_{ZZ} and Q_{XX} components. The experimental second moment component relative to a chosen molecular origin, ($Q_{XX} = -55.53$, $Q_{ZZ} = -63.88$) shows a weaker charge expansion than in the free molecule ($Q_{XX} = -53.63$, $Q_{ZZ} = -51.53$) while the positive Q_{XZ}'s indicate a similar contraction in the $(\vec{X}+\vec{Z})$ direction (orientations in the molecular frame given in Figure 5) for both the free molecule and the molecule in the crystal state. On the other hand the same electronic delocalization in the $(\vec{X}+\vec{Z})$ direction is being observed in the molecular plane for molecules in both states.

Table 7. Components of the molecular quadrupole moment of the charge distribution(e.Å^2) from theoretical calculationsand experimental electron density study.

QuadrupoleMoments	X-Ray Experiment	*AbInitio*DFT(6-31G)
Q_{XX}	−55.532	−53.632
Q_{YY}	−53.129	−53.777
Q_{ZZ}	−63.886	−51.536

Table 7.*Cont.*

Quadrupole Moments	X-Ray Experiment	*Ab Initio* DFT(6-31G)
Q_{XY}	−1.825	0.964
Q_{XZ}	3.878	0.002
Q_{YZ}	−1.755	−0.001

4.5. Frontier Molecular Orbital Analysis

Molecular orbitals (HOMO-LUMO) and their properties such as energy are very useful for physicist and chemists and are very important parameters for quantum chemistry. This is also used by the frontier electron density for predicting the most reactive position in π-electron systems and also explains several types of reaction in conjugated system [19]. The conjugated molecules are characterized by a small highest occupied molecular orbital- lowest unoccupied molecular orbital (HOMO-LUMO) separation. Both the highest occupied molecular orbital and lowest unoccupied molecular orbital are the main orbitals which take part in chemical stability. The HOMO represents the ability to donate an electron, LUMO as an electron acceptor, represents the ability to obtain an electron. The HOMO and LUMO energy calculated by B3LYP/6-311++G(d,p) method is shown below. This electronic absorption corresponds to the transition from the ground to the first excited state and is mainly described by one electron excitation from the highest occupied molecular orbital to the lowest unoccupied molecular orbital. While the energy of the HOMO describe the ionization potential, LUMO energy is concerned by the electron affinityEnergy difference between HOMO and LUMO orbital is called as energy gap which is an important stability for structuresand is calculated as:

HOMO energy=−0.264 au
LUMO energy=−0.106 au
HOMO-LUMO energy gap =−0.158 au

It has been shown that calculated energy gap between HOMO and LUMO can be very useful to prove the activity from intramolecular charge transfer [20].

4.6. Electrostatic Potential

In order to grasp the molecular interactions, the molecular electrostatic potential (MEP) is used. The molecular electrostatic potential is the potential that a unit positive charge would experience at any point surrounding the molecule due to the electron density distribution in the molecule. The electrostatic potential is considered predictive of chemical reactivity because regions of negative potential are expected to be sites of protonation and nucleophilicattack,while regions of positive potential may indicate electrophilic sites.The distribution of the electrostatic potential for the molecule in the crystal was calculated from Equation (7):

$$\Phi(r) = \int \frac{\rho_{total}(r)}{|r - r'|} dr \tag{7}$$

where ρ_{total} represents both the nuclear and the electronic charge. The integration is over the molecular volume, and r' represents the atomic position relative to same origin.The integration includes the atoms of only one molecule and therefore does not include directly the effects of charge distribution of the molecules.

Figure 6 shows the experiment and theoretical maps of the electrostatic potential distribution in the plane of the base ring.We are used the Density Functional Theory at B3LYP level of theory at 6-31G* to describe the theoretical electrostatic potential map. Figure 7 is the same representation in 3D dimensions of the theoretical electrostatic potential map.The extension of the positive electrostatic potential around the C-H group and the regions of negative electrostatic potential around the nitro and hydroxyl group gives same conclusion about the nature of the intramolecular charge transfer as found by the orientation of the molecular dipole moment.

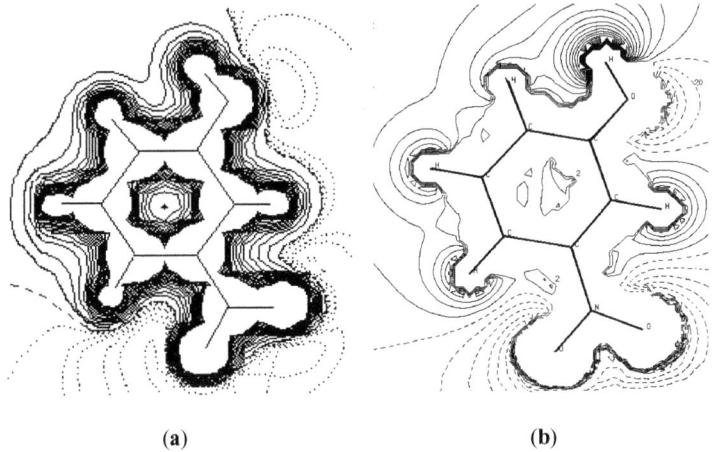

(a)   (b)

Figure 6. The electrostatic potential maps around the molecule. The section is in the plane of the ring atoms. (a) Experimental (contours are at $0.05 e\text{Å}^{-1}$). (b) Theoreticalusing the Density Functional Theory at B3LYP level of theory at 6-31G* (contours are at $0.025\ e\text{Å}^{-1}$). Zero and negative contours are dashed lines($1 e\text{Å}^{-1} = 332.1 \text{kcal·mol}^{-1}$).

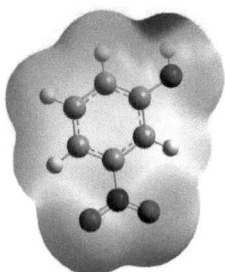

Figure 7. 3D-representation of the electrostatic potential around the molecule using the Density Functional Theory at B3LYP level of theory at 6-31G*

The potential of the *m*-nitrophenolmolecule has been calculated from the experimental electron density distribution by the multipolar method using the X-ray diffraction data. The comparison of the experimental potential in a crystal and the theoretical potential for an isolated molecule is an excellent test for high quality descriptive model for the electron charge density distribution from X-ray diffraction experiment.

5. Conclusions

In this article, we have dealt with the salient features of the electronic charge density distribution in molecular solids obtained by both theory and experiment. This study has obtained good accurate results on the structure and electron charge density which back the experimental results for the electron charge density distribution.

The general conclusion fromthe estimation of the dipolar moments and the electrostatic potential of the *m*-nitrophenolmolecule in the both experimental and theoreticalstudy is that the region of the nitro and hydroxyl groups is electronegative and the C-H group region is electropositive. These results could be used to explain the existence of the polymorphism in *m*-nitrophenolcompounds, if they were completed by the study of the nature and the energy of the molecular interaction by the X-ray diffraction of the both polymorphic of m-NPH.

References

1. *Panadares, F.; Ungaretti, L.; Coda, A. The crystal structure of a monoclinic phase of m-nitrophenol.*ActaCryst.B*1975*, 31, 2671–2675.
2. Wojcik, G.; Toupet, L. The inter-and intramolecular charge transfer along the polymeric chain of hydrogen-bonded molecules in two crystal forms of m-nitrophenol. *Mol. Cryst. LiquidCryst.* **1993**, *229*, 153–159.
3. Hamzaoui,F.; Baert, F.; Wojcik, G. Electron-density study of *m*-nitrophenol in theorthorhombic structure. *ActaCryst. B*1996, *52*, 159–164.
4. Wojcik, G.; Marqueton, Y. The phase transition of m-nitrophenol. *Mol. Cryst. LiquidCryst.* **1989**, *168*, 247–254.
5. Hamzaoui,F.;Drissi, M.; Chouaih, A.; Lagant, P.; Vergoten, G. Electron Charge Density Distribution from X-ray diffraction Study of the M-Nitrophenol compound in the monoclinic form. *Int. J. Mol. Sci.*2007, *8*, 103–115.
6. Wojcik,G.;Mossakowska, G. Polymorphs of p-nitrophenol as studied by variable temperature X-ray diffraction and calorimetry: Comparison with m-nitrophenol.*ActaCryst.*2006,*B62*, 143–152.
7. Wojcik, G. Intermolecular interactions leading to crystal polymorphism of organic compounds.
    X-ray diffraction and quantum chemical studies of *para*- and *meta*-Nitrophenol. *Ser. Khim.* **2007**,*48*, 123–127.
8. Jeffrey, G.A.; Cruickshank, D.W.J. Molecular structure determination by X-ray crystal analysis: Modern methods and their accuracy. *Q. Rev. Chem.Soc.* **1953**, *7*, 335–376.
9. Coppens, P.;Volkov, A.The interplay between experiment and theory in charge-density analysis. *ActaCryst.*2004,*A60*, 357–364.
10. Hansen, N.K.; Coppens, P. Testing aspherical atom refinements on small-molecule datasets.
    *ActaCryst. A*1978, 34, 909–921.
11. Jelsch, C.;Guillot,B.;Lagoutte, A.;Lecomte, C. Advances in protein and small-molecules charge-density refinement methods using MoPro.*J. Appl. Crystallogr.*2005, *38*, 38–54.
12. Volkov, A.; Abramov, Y.; Coppens, P.; Gatti, C. On the origin of topological differences between experimental and theoretical crystal charge densities. *ActaCryst. A* **2000**, *56*, 332–339.
13. Volkov, A.; Gatti, C.; Abramov, Y.; Coppens, P. Evaluation of net atomic charges and atomic and molecular electrostatic moments through topological analysis of the experimental charge density.*ActaCryst. A* **2000**, *56*, 252–258.
14. Coppens, P.*X-ray Charge Densities and Chemical Bonding*;Oxford University Press:New York, USA, 1997.
15. *Hehre, W.J.; Schleyer, R.R.; Pople, J.A.AbInitio Molecular Orbital Theory; John Wiley &Sons:Hoboken, NJ, USA, 1986.*
16. *Frisch, J.; Trucks, G.W.; Schlegel, H.B.; Scuseria, G.E.; Robb, M.A.; Cheeseman, J.R.; Montgomery, J.A.,Jr.;Vreven, T.; Kudin, K.N.; Burant, J.C.; et al.Gaussian 03, Revision A.1; Gaussian, Inc.: Pittsburgh, PA, USA,2003.*
17. Schaftenaar, G.;Noordik, J.H.Molden: A pre- and post- processing program for molecular and electronic structures.*J.Comput.-AidedMol. Des.*2000, *14*, 123–134.
18. Blessing, R.H. DREAD—data reduction and error analysis for single-crystaldiffractometer data. *J. Appl. Cryst.* **1989**, *22*, 396–397.